Gonna Trouble the Water

GONNA TROUBLE THE WATER

Ecojustice, Water, and Environmental Racism

Edited by
Miguel A. De La Torre

Foreword by
Governer Bill Ritter

the pilgrim press
since 1640

The Pilgrim Press
1300 East 9th Street
Cleveland, Ohio 44114
thepilgrimpress.com

Published 2021.

Printed on acid-free paper.

Library of Congress Cataloging-in-Publication Data on file
Library of Congress Control Number: 2020947482
ISBN 978-0-8298-2169-7 (alk . paper)
ISBN 978-0-8298-2171-0 (ebook)

Printed in The United States of America.

to

Nick R.

Contents

Foreword

Governor Bill Ritter

When the chapters for this book were presented by the authors at a forum at the Iliff School of Theology, it was my privilege to be present and to be invited to present as well. Although I run a clean energy policy think tank at Colorado State University, I learned much about water on my family's dry land wheat farm, then during my three-year stay in the Western Province of Zambia running a nutrition center, and finally during my time as Colorado's governor. With that said, when I observed the presentations, and I read the chapters in this book, I was in awe. I was compelled to consider water, and the many intersections explored here, in ways I never had before.

There is little doubt that water, like blood, is life-giving. And owing to its precious nature, water has become one of many natural resources in our ecology that has been bought and sold, too often not on equitable terms, used as a lever in uneven negotiations, poisoned to the detriment of the marginalized, or just downright stolen. And that is to name only a few of the ways that water has been mismanaged and mistreated. In the most elegant and eloquent of terms, these authors look at the deep and abiding intersections of water with race, with indigenous peoples, with development, with ecology generally, and with a host of other issues as well. And, at every turn, we are required to consider issues of justice generally, and ecological justice specifically.

I believe that most people in America, and throughout much of the globe, care a great deal about justice, about fair and equal treatment. But too often, justice concerns are thought about in overly simplistic terms—no one should be the victim of a crime, violent and non-violent alike. No one should be punished for the bad conduct of another. There should recompense if someone defrauds another, or deceives them in a business transaction, or injures them with tortious conduct.

In fact, in the United States and around the globe, we have institutionalized a rule of law and a court system for just these reasons. But what happens when we think more deeply about justice, injustice, and the natural world, when we think about natural resources, and perhaps most important among them, when we think about *water*.

The chapters of this book will answer many of those questions from the perspectives of the respective authors, and there is nothing simplistic about it. Their writings revolve around ecological justice, but at a deeper level they make me weep for indigenous peoples who were stewards of water until it was taken from them, for marginalized communities who were an afterthought where access to clean water was concerned. They make me weep for land that once was fertile and fecund—and now sits parched and sterile. My hope is that we will someday live in a world where ecological justice will be not a product of conflict resolution, but at the forefront of every human's mind. These chapters and these authors can help us find our way there.

Introduction

Miguel A. De La Torre

The forty-fifth president of the United States appears to have a distressing and disconcerting obsession with water—specifically, toilet water. On the eve when the US House of Representatives voted, mainly along party lines, to impeach Donald Trump on two articles—abuse of power and obstruction of Congress—the president spoke at a rally in Battle Creek, Michigan (December 18, 2019), one would have thought he would have professed his innocence or laid out his defense against impeachment. Instead he spent a good portion of his speech waging an ongoing feud against bathroom fixtures with water-conserving features. He was perturbed that bathroom fixtures, specifically toilets, now require ten flushes as opposed to just one. According to the president: "Ten times, right? Ten times. [He then made a flushing motion with what appeared to be a flushing sound, "bah, bop"] Not me of course, not me, but you. You." In Trump's mind, there exist places where there is so much water we simply don't know what to do with it.[1] So on the night of his impeachment, a historical event that has only occurred two other times in this nation's history, the president of the United States is telling a rally of supporters that he does not defecate much, but they do—so much that it requires ten flushes to finally remove all the manure because of the inefficiency of water-efficient toilets.

While his fevered rant against water conservation might appear humorous if expressed by one's favorite drunk uncle during a family gathering, it becomes life-threatening when spoken by the so-called leader of the free world. His directives to the Environmental Protection Agency (EPA) to adjust its policies in order to deregulate toilets may appear harmless and probably will go nowhere;[2] nevertheless, his other actions concerning water encompass a greater threat. As president, and not just some delusional baby boomer, he has the power to impose earth-destructing policies upon the world based on his conception of water. Since taking office, while ignoring science-based reports on the stresses existing on the world's water supplies, he has completely rolled back five federal regulations specifically dealing in safeguarding our water supply and is in process of rolling back two additional regulations.[3] For example, on September 12, 2019, the Trump administration redefined and restricted which waterways fell under the protection of the federal Clean Water Act, which led to the loss of federal protection for 51 percent of wetland, and 18 percent of streams throughout the United States. An immediate effect is that polluters will no longer be required to obtain a permit to discharge potentially harmful substances. These wetlands and streams serve as the headwaters for lakes and rivers acting as filters keeping pollutants and farm fertilizer runoff from flowing downstream, thus providing millions of people clean drinking water.[4]

According to the EPA's Scientific Advisory Board, a top panel of government-appointed scientists responsible for evaluating the scientific integrity of public policies, President Trump's proposal to rewrite the Obama-era regulation of waterways "neglects established science," by "failing to acknowledge watershed systems." The forty-one scientists, many of whom were hand-selected by the Trump administration, found "no scientific justification" for excluding certain bodies of water from protection under the new regulations.[5]

Regardless if some future 46 attempts to undo the damage that 45 wrought, the injury to the nation's fragile water systems may well be irreversible. Trump, along with many Euro-Americans, have a misconception concerning water, one where it is reduced to a plentiful commodity exploitable for gain or convenience. Water exists to be used, misused, and abused by those who can profit from its availability. But even those within the dominant culture, those who are more aware of the environmental degradation occurring, still gaze upon water from a position of privilege. Yes, they may be concerned with the sustainability of our planet in general, and water, specifically; nevertheless, their concern is anchored in a Eurocentric worldview that remains irreconcilable with the worldviews emerging from the margins of society. Eurocentric environmentalists mainly profess a commitment to clean air and water or the protection of endangered species. No doubt these are noble and necessary pursuits. But gazing upon water from a position of power and privilege results in a radically different analytical assessment than when conducted from the underside, from the margins, from the perspective of the disenfranchised. If this book had been written by predominately white scholars instead of scholars from or in solidarity with racially and ethnically marginalized communities, the focus on water would have been on the importance of clean water, how we are falling short of this goal, and what actions we can individually or collectively engage which might roll back the devastation being wrought due to neoliberal greed. Again, a worthy discussion in which to engage. But an intention was instead made to gather scholars from or in solidarity with marginalized communities and to gaze upon water through *their* eyes, though *their* worldview. Such a perspective places the accent upon different considerations. According to Robert Bullard:

> The environmental-equity movement is an extension of the social justice movement. Environmentalists may be concerned about clean air, but may have opposing views on the construction of low-income housing in white, middle-class, suburban neighborhoods. . . . It is not surprising that mainstream environmental organizations have not been active on issues that disproportionately impact minority communities. . . . Yet, minorities are the ones accused of being ill-informed, unconcerned, and inactive on environmental issues.[6]

Unlike most books written from the dominant culture concerning the environment in general, water specifically, the contributors of this text, predominately from minoritized communities, sought to emphasize the environmental quality of life where they are forced to live as opposed to the overall white community. Soon, it became clear a link exists between the water pollution they experience and their minoritized status. Unfortunately, few white environmentalist colleagues consider or examine this phenotype link. Consequently, they fail to grasp how water pollution disproportionately impacts nonwhite communities. The failure of the environmental justice movement to consider the inherent racism in most environmental discourses relegates those on their margins to greater ecological devastation that, unsurprisingly, often leads to greater health risks.

This book seeks to break with the normative environmental conversation concerning water by explicitly linking the liquid substance on which all life depends with racism and ethnic discrimination. To that end, the different contributors of this text gathered on October 24–25, 2019, at the Iliff School of Theology in Denver, Colorado, to discuss among themselves the intersection of water and environmental racism. We who gathered consisted of scientists, journalists, international lawyers, activists, government officials—and, of course, religion scholars. We represented multiple racial and ethnic perspectives, seeking commonality while remaining cognizant of our differences. We listened to and debated with each other, always pushing ourselves not to simply mimic the discourse already taking place among our Eurocentric colleagues. After days of wrestling with the issue, we discovered our presentations fell into three categories.

The first saw water as sacred. From the depth of our communities' worldviews, unlike the Eurocentric concept of water as a commodity to be exploited, we saw water as a sacred living entity possessing its own rights and dignity. This sacred aspect to water is not only responsible for giving life, but also liberation. To secularize water is to imprison those who rely on its life giving and spiritual enhancing qualities. The first four chapters of the book wrestle with this understanding of the character of water. The second category understood access to water as a human right. Not only does water possesses its own rights, all that is living—not just humans—has a right to enter into a relationship with water. To deny water is to deny life. To pollute water is to pollute life. And finally, the last category examined water as a tool to abuse, a weapon against minoritized communities. This section examines the weaponization of water in order to control non-Eurocentric bodies. That which gives life is here re-signified to deliver death. As this introduction indicates, this book is more than simply an environmental study of water and its degradation. This project is an attempt to approach the intersection of water and racism by mooring the discourse in the voices and worldviews of the marginalized. To that end, we hope to introduce a somewhat different perspective to the overall discourse from its underside.

Notes

1. Brittany Shammas, "Trump Was Impeached. But Dishwashers That Go 'Boom' Are on His Mind," *Washington Post,* December 19, 2019.

2. Justin Sink and Mario Parker, "Trump Orders Toilet Rule Review Over Low-Flow Flushing," *Washington Post,* December 7, 2019.

3. Nadja Popovich, Livia Albeck-Ripka, and Kendra Pierre-Louis, "85 Environmental Rules Being Rolled Back Under Trump," *New York Times,* September 12, 2019.

4. Lisa Friedman and Coral Davenport, "Trump Administration Rolls Back Clean Water Protections," *New York Times,* September 12, 2019.

5. Coral Davenport and Lisa Friedman, "EPA Policies Scorn Science, Panel Reports," *New York Times,* January 1, 2020.

6. Robert D. Bullard, *Dumping in Dixie: Race, Class, and Environmental Quality,* 2nd ed. (Boulder: Westview, 1994), 128–29.

WATER AS SACRED

CHAPTER ONE

Water—A Living Spirit

Miguel A. De La Torre

That the Iliff School of Theology held its inaugural Ecojustice conference on the intersection of water and environmental racism during the fall of 2019 is apropos; after all, Iliff, founded in 1892, exists because of the commodification of the water. The school's namesake, John Wesley Iliff (1831–78), greatly profited from the imposition of a Eurocentric worldview that saw natural resources, like water, as a commodity to be exploited for personal economic gains rather than a living sacred entity. While fortune seekers, known as the "Fifty-niners" traversed what they observed to be the wastelands of the Colorado plains to reach the Rocky Mountains where they prospected for gold, Iliff—the man, not the school—saw an opportunity. Originally born in McLuney, Ohio, he headed west with $500 in his pocket, arriving in Denver in 1859. Rather than mining for riches, he found riches in mining miners. From his general store off Larimer between F and G Streets, Iliff provided the goods miners needed to stake their claims. By 1861, he began buying land along the South Platte River for his emerging cattle ranch. He discovered undernourished cattle, cheaply bought, could easily be fattened if allowed to graze, increasing their economic worth. After 1868, Iliff was providing beef to the Union Pacific and Colorado Central Railroad crews, army forts, the Cheyenne boomtown, the Chicago Stockyards and several Indian reservations.[1]

His holdings were so vast that his cattle grazed over 650,000 acres. He did not own all this land; what he owned was the water rights. Manipulating the Homestead Act (contrary to the original intent of the US Congress), Iliff was able to control water rights between Julesburg and Denver. Owning the water is what made the land useful for other ranchers and farmers.[2] Because Iliff was the first person to divert water for a beneficial use, he acquired sole rights to the water. Taking advantage of the doctrine of prior appropriation, which was based on the legal concept *qui prior est tempore potior est jure* (the one who is

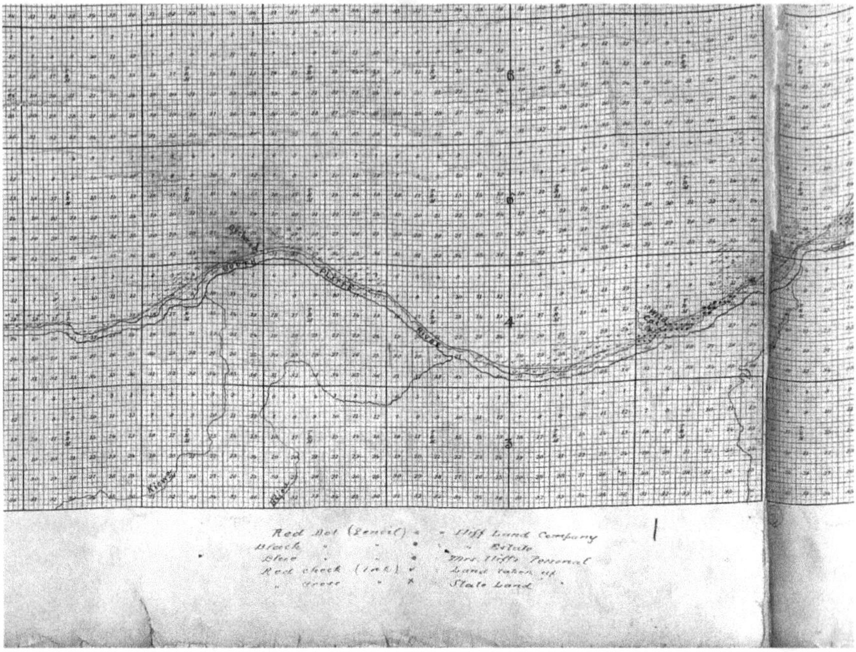

FIGURE 1. Hand-drawn cloth map of land John Wesley Iliff controlled along the South Platte River (Margaret E. Scheve Archives and Special Collections, Iliff School of Theology, Denver. Photo by Erin Shafer)

first in time is first in right), Iliff gained perpetual legal rights to the South Platte's flow of water along with its streams and offshoots.[3]

A monopoly on water meant that no one could homestead between his lands without purchasing water directly from him. Water is what made land—private and public—valuable, and because he owned the water, he was able to create a cattle empire. His cattle grazed on so-called public land (land still belonging to the indigenous people) because he provided the water that made the land grazable. Iliff's raising of cattle would contribute to the devastation of the local bison. He was among those who within two decades participated in driving the buffalo to near extinction, reducing the numbers south of Canada to 325 and replacing them with about 5.5 million cattle, allowing him to be among the first to earn the sobriquet "cattle king."[4]

An Indigenous Worldview

Iliff has been portrayed as a visionary, a seer of the future. He gazed upon grazable empty land not being maximized for profit and, by taking control of flowing water, made a fortune that has directly benefited the author of this chapter, who serves as a tenured professor at the institution his family founded. Iliff's fulfilled a manifest destiny, a worldview that credited his Methodist God for richly blessing him for his faithfulness. But there exists another worldview that questions whether anyone can own water; for water is alive and upon it all life depends, not just humans who eat the cattle, but the cattle who eats the grass, and the grass itself. Smohalla of the Wanapum, a different kind of seer, asserted in 1890:

> You ask me to plow the ground. Shall I take a knife and tear my mother's bosom? Then when I die she will not take me to her bosom to rest. You ask me to dig for stone. Shall I dig under her skin for bones? Then when I die I cannot enter her body to be born again. You ask me to cut grass and make hay and sell it, and be rich like white men. How dare I cut off my mother's hair.[5]

Cognizant of the violence in placing words into someone else's mouth, I nevertheless wonder if Smohalla would have also agreed with the imagery that hoarding water for the gains of one person would be similar to draining the blood from one's mother's veins?

Indigenous scholar George "Tink" Tinker reminded me during a conversation that within the North American indigenous worldview, the river, the spring, and the lake all have their own living being. In his own tradition, the Osage word for river is *ni*. Every *ni*, according to him, is a person, alive, and a close relative on whom our lives depend. As such, they are due the respect granted to any human being. Contrary to the Eurocentric worldview embodied by John Wesley Iliff, which saw bodies of water as commodities needing to be exploited for personal enrichment, indigenous people recognized these bodies of water as life. For example, the Overhill Cherokee of what is today eastern Tennessee knew the Little Tennessee River as *Yunwi Gamahida,* translated as the long Man, "a living being who granted energy and spoke to those who knew how to listen."[6] The Taos Pueblo Community, in what is today called New Mexico, understood that their earliest ancestors emerged from the cold waters of *Ba Whyea* (Blue Lake); and into these waters "the dead return, to become spirits, known as 'the ones who always believe' [where] they merge with cloud people, the rain-bringing Pueblo Indian spirits known as *Kachinas.*"[7]

In what would become Central America, the Aztec cosmology includes *Chalchiuhtlicue* (She of the jade skirt), the goddess of rivers, lakes, streams, and other freshwaters; wife (or sister) to *Tlaloc*, the god of rain. She is associated with fertility and is responsible for bringing forth maize, the life source of the people. There is also a goddess of saltwater, *Huixtocihuatl*, a fertility deity. The indigenous worldview of the Américas is not simply some quaint way of understanding water; rather, such an understanding resonates with other indigenous groups throughout the world. Take, for example, the Yoruba people of what today is called Nigeria. They, too, see rivers and streams as a living entity called Oshún, patron of love and eros. Oceans are under the domain of Yemayá, who is associated with womanhood and motherhood. Inle is associated with the fertility of water to produce crops. And Olocun, considered the queen of the ocean's depth, is responsible for the worldwide flood.[8] Likewise, we can look to Asia where China's Yellow River, one of the world's major water arteries, deity is Hebo, who like the river is "benevolent [yet] greedy, unpredictable and dangerously destructive."[9] Or we can look to the sacred Ganges River of India, the holiest of all Hindu rivers, synonymous with the goddess Gaṅgā. Her waters are the "liquid embodiment of *śakti* [female principle of divine energy] as well as the sustaining immortal fluid (*amrta*) of mother's milk." To bathe in her waters facilitates liberation from the life-and-death cycle (*moksa*).[10] Even European indigenous peoples, prior to the imposition of Christianity, recognized the sacredness of water, whether they be the Celtic people of Ireland who knew the River Boyne to be the goddess Boann; the Welsh who looked to Llyr, god of the sea; or the Gaul who recognize the River Seine as the goddess Sequana.

If indeed the worldview of the majority of the earth's indigenous peoples are closer to reality than the Eurocentric neoliberal worldview which is rooted in the colonizing capitalist philosophical trend rooted in Christianity and the so-called Age of Enlightenment, then to own water becomes a form of enslavement, not just of the living entity associated with water, but upon those who depend on the water. To impose the Eurocentric worldview upon the planet's inhabitants extends a global oppression which enriches a minority of the global population while creating the illusion of white supremacy which confuses unearned riches from natural resources with advanced civilization. Once water is stripped of its spiritual entity and reduced to a commodity, it can be manipulated to benefit the global minority at the expense of the global majority.

A Eurocentric Worldview

Rather than recognizing water as a living entity, Western religions like Christianity have focused on the relationship established between humans and what they perceive to be the one true God. Based on this human-divine relationship, intra-human relationships are established, which usually are portrayed as positive for those who share one's beliefs or negative toward those with a different worldview faith tradition. Missing from the Christianity of the colonizers has been any interconnectedness with a collective or communal spirituality linked to water. Emphasis, among many Protestants, usually is tied to some "personal" relationship with Jesus. Little attention is given to the relationship between humans and other spiritual entities.

If creation stories describe humanity's appointment as stewards of the earth's resources, then human beings as caretakers are called to protect, preserve, and safeguard these living resources so all can benefit and enjoy its fruits. Creation as gift means all that is living has basic rights, one of which is that no person or group has the right to hoard the earth's resources. Hoarding the earth's resources upsets the delicate balance between life and those dependent on said resources to sustain life. Ironically, Christianity has encouraged the destruction of creation. The first creation story found in the Hebrew Bible ends with: "And God said to [human beings], 'Be fruitful and multiply, and fill the earth and subdue it; and have dominion over the fish of the sea and over the birds of the air and over every living thing that moves upon the earth" (Gen 1:28). Biblical passages such as these led to and justified human domination and ownership of nature, specifically, the water and the land. Christians have come to read this verse as permission to use and abuse creation, similar to how the ancient rulers once used and abused their subjects. Man (and here I mean those who are men) are ordained to rule. This understanding of human (specifically, the male) occupying the pinnacle of creation has been echoed throughout Christian history, most recently by Pope John Paul II who stated, "Everything in creation is ordered to man and everything is made subject to him."[11]

Western Christianity's understanding of stewardship and domination as subjugating nature contributes to the present ecological challenges that humanity faces. The belief that the destiny of human beings is to reside with God in heaven and that the earth is but a place of sojourn until we reach our celestial destiny, has encouraged—at the very least—neglect of our environment. The greatest threat to the environment comes from those who hold a view of the end of time (an "eschatological" view) where the destruction of the earth is welcomed for it indicates Jesus's "second coming" when he raptures (takes away from the earth) those destined to be saved. If the world ends in a conflagration and such an end is close at hand, why then worry about the environment?[12]

Contrary to these fundamental Christian views, most indigenous religions from the Global South maintain a sacred respect for creation, a respect lost and historically abused by many Western religions like Christianity. Indigenous traditions that understand water as living entities, if not deities, remind us that within the circle of creation, all are equal in value to the Creator. Or as George "Tink" Tinker writes: "A chief is not valued above the people; nor are two-legged valued above the animal nations, the birds, or even trees and rocks."[13] Even the stones cry out the praises of creation. Human relationships to creation become a matter of life and death, balancing one's needs and place within the world while seeking to preserve the world for one's descendants who will live "seven generations from now." While one takes from the plenty of creation, something must always be returned to maintain balance.[14] If we are going to appropriate from nature, we must reciprocate to maintain a mutual relationship. Aruna Gnanadason, a scholar from India, rejects the dualism intrinsic in Western theological thought, which is based upon the opposition between the spirit and flesh, men and women, or mind and body. Instead, she insists on a cosmology that affirms the interdependence and harmony of all life forms.[15] It is not a choice between humans and nature as usually framed when debating political policies and decisions that negatively impact the environment, but rather a both/and, where humans are but one of the many life entities that is one with nature. Contrary to Eurocentric either/or binary opposition, Asian theologians like Kwok Pui-lan instead emphasize the balance of "heaven and earth, yang and ying, sun and moon, and father and mother" where they are "complementally, mutually reinforcing and interplaying with one other."[16]

Stewardship of the earth's resources requires creating a harmonious relationship with nature, as with other human beings. Water is not to be enslaved, sold, or mistreated. To do so creates pollution, lowers life expectancy, and becomes the source of illnesses and diseases for those living close to abused waterways—those who are predominately non-European. The exploitation of the earth's water, and the exploitation of the earth's disenfranchised are interconnected, making it difficult, if not impossible, to speak of one without mentioning the other.[17] Brazilian liberationist theologian Leonardo Boff has connected the cry of the oppressed with the very cry of the earth. He insists that the logic and justification that lead the powerful and privileged to exploit and subjugate the world's marginalized are the same logic and justification that plunder the earth's wealth and lead to its devastation.[18] Limiting the risk of the industrialized North is directly linked to increasing the risk of the Global South.

Take, for example, the musing of white liberal political leaders at the World Bank who saw the environmental degradation of the planet as a Eurocentric right rooted in the ownership of the earth by those who deem themselves superior. Lawrence H. Summers was the World Bank's chief economist and vice president for development economics from 1990 until 1993. When he left the World Bank, he went to serve the Clinton administration as undersecretary of the US Treasury Department, eventually becoming secretary of the treasury from 1999 until 2001. Afterwards, he served as the president of Harvard University from 2001 to 2006. President Obama would tap him in 2009 to serve as the director of the White House National Economic Council.

While vice-president at the World Bank, Summers came to be known as the Bank's "high priest" responsible for supervising all economic publications, including the pacesetting *World Development Report*. Commenting on the topic of "dirty industries," the theme discussed by one of the publications he oversaw, he wrote a memo to six highly placed colleagues. This private memo was eventually leaked to the press. In it, he wrote:

> Just between you and me, shouldn't the World Bank be encouraging more migration of the dirty industries to LDCs (Less Developed Countries)? . . . The measure of the costs of health impairing pollution depends on the foregone earnings from increased morbidity and mortality. From this point of view a given amount of health impairing pollution should be done in the country with the lowest

> cost, which will be the country with the lowest wages. I think the economic logic behind dumping a load of toxic waste in the lowest wage country is impeccable and we should face up to that. . . . I've always thought that underpopulated countries in Africa are vastly underpolluted, their air quality is probably vastly inefficiently low compared to Los Angeles or Mexico City.[19]

In Summer's mind, life is not priceless. All that has life, whether it be people, animals, plants, rivers, land, air, or rocks, have a monetary value. Ignoring their spiritual entities, Summers concludes that if "a load of toxic waste" is dumped in a rich country, it would cause the infirmity and death of high wage earners with normally long life expectancies. According to Summers's own estimates: $20,000 per year of potential earnings for a forty-year-old with an estimated twenty-five more years of productivity can contribute $500,000 to the global economy. By contrast, if the "load of toxic waste" is dumped in a poor country with a GNP of $360 and average life expectancy of fifty-five years, the contribution of a worker to the global economy could be calculated at a measly $5,400.[20] Such an approach reduces all living entities to an economic value. Not surprisingly, people of Eurocentric origins with economic privilege are always worth more than the planet's disenfranchised. The sacredness of profits replaces the sacredness of life. Death-causing pollutants, therefore, should be "dumped" in poor countries, creating a new pattern of implementing colonialism and imperialism.[21]

Implementation of a Eurocentric Worldview

We can see how the implementation of Summers's Eurocentric capitalist worldview materialized in Savar, Bangladesh, where a toxic stench rises off a polluted canal in which garment factories dump their wastewater. These factories produce cheap clothing for Western department stores like Walmart, J. C. Penney, and H&M. In so doing, they have contributed to the ecological devastation of the area. Rice paddies are inundated with toxic wastewater, fish stocks are decimated, coconut trees have stopped producing fruit, and smaller waterways are filled with sand and garbage to sell off as land for further housing or factory development. The situation is so horrific that children of garment workers who attend Genda Government Primary School are often dizzy and lightheaded with some retching during class. Unsurprisingly, they are unable to concentrate on their studies. To make matters worse, Bangladesh is acutely vulnerable to global climate change. Millions of inhabitants would be displaced and crop yields would significantly drop with the onslaught of changing weather patterns and rising sea levels.[22]

This type of ecological destruction is not limited to those living beyond the borders of the United States. Environmental racism, understood as the link between the degradation of the environment and the racial composition of the areas where said degradation occurs, are just as prevalent among communities of color within the borders of the United States. Race, according to a growing body of empirical evidence, continues to be the most significant variable in determining the location of commercial, industrial, and military hazardous-waste sites. As ethicist Christina Zenner reminds us: "Water is currently among the most visible sites of differential, systemic privilege. It is through the substance of water that race, class, and geography in Flint and Standing Rock are revealed to be intimately and dangerously related."[23]

Using US census data from 2000, people of color represent 56 percent of the population who live less than 1.8 miles from one of the 413 commercial waste facilities.[24] The poorer the community, the greater the risk of environmental abuse, because those economically privileged are always able to move away from such sites, a privilege not available to the poor, who are mostly people of color. Between 1999 and 2009, the National Academy of Science produced five environmental justice reports showing that "low-income and people of color are exposed to higher levels of pollution than the rest of the nation and that these same populations experience certain diseases in greater number than more affluent White communities."[25]

Womanist ethicist Emilie Townes has said that the effects of toxic waste on the lives of people of color who are relegated due to their poverty to live on ecologically hazardous lands are akin to a contemporary version of lynching a whole people.[26] The Global North is lynching the Global South with watery ropes. Award-winning poet Alice Walker shares a similar sentiment when she writes, "Earth itself has become the n*gg*r of the world." She goes on to elaborate:

> It is perceived, ironically, as other, as alien, evil, and threatening by those who are finding they cannot draw a healthful breath without

> its cooperation. While the Earth is poisoned, everything it supports is poisoned. While the earth is enslaved, none of us is free. While the Earth is a "n*gg*r," it has no choice but to think of us as all as Wasichus [first Europeans to meet the Lakota]. While it is "treated like dirt," so are we.[27]

Water, or the earth itself, may be a living entity; however, the white Eurocentric neoliberal economic worldview has reduced these living elements to a creature to be used, misused, and abused for profit by the chosen few.

When those who are racially and ethnically marginalized compare the environmental quality of life of where they live with that of the larger white society, it becomes crystal clear that a link exists between polluted sites and disenfranchisement. Few white environmentalists seriously consider this link and consequently fail to fully understand environmental issues. The failure of the environmental justice movement to come to terms with the inherent racism which relegates those on the margins to the greatest ecological health risks prevents fostering a truly global, holistic approach to the environment. Continuing to mask environmental racism limits, if not frustrates, any attempt or hope for the liberation of humans, creaturekind, and the living elements of nature alike. In the final analysis, John Wesley Iliff embodied a philosophy that led to the winning of the West. About a century later, Lawrence H. Summers, the former chief economist of the World Bank, seeks to expand this philosophy to win the whole world, enunciating the neoliberal Eurocentric worldview when he said, "Promoting development is the best way to protect the environment."[28] Those from the margins of society who are experiencing "a load of toxic waste" dumped upon them in the name of development might differ.

Notes

1. Ira J. Taylor Library Archives, *The Iliff Family Papers Biographical Sketches* (Denver: Marshall Edison, 2002).

2. Thomas J. Noel and Duane A. Smith, *Colorado: The Highest State,* 2nd ed. (Boulder: University Press of Colorado, 2011), 141–42.

3. Wells A. Hutchins, *Water Rights Law in the Nineteen Western States* (Clark, NJ: The Lawbook Exchange, 2004) 1:396–400.

4. Christopher Knowlton, *Cattle Kingdom: The Hidden History of the Cowboy West* (New York: Houghton Mifflin, 2017), 6–7.

5. James Mooney, "The Ghost-Dance Religion and the Sioux Outbreak of 1890," in *Fourteenth Annual Report of the Bureau of Ethnology to the Secretary of the Smithsonian Institution 1892–93,* Part 2 (Washington, DC: Government Printing Office, 1896), 721.

6. Peter Nabokov, *Where the Lightning Strikes: The Lives of American Indian Sacred Places* (New York: Penguin, 2006), 53–54.

7. Ibid., 73–74.

8. Miguel A. De La Torre, *Santería: The Beliefs and Rituals of a Growing Religion in America* (Grand Rapids, MI: Eerdmans, 2004), 36–37, 48–49.

9. Shan hai jing, *A Chinese Bestiary: Strange Creatures from the Guideways through Mountains and Seas,* ed. and trans. Richard E. Strassberg (Berkeley: University of California Press, 2002), 202.

10. Diana L. Eck. "GAṄGĀ: The Goddess Ganges in Hindu Sacred Geography," in *Devī: Goddesses of India,* ed. John S. Hawley and Donna M. Wulff (Delhi: Motilal Banarsidass, 1998), 137–38.

11. Pope John Paul II, *The Gospel of Life* (New York: Random House, 1995), 61.

12. Miguel A. De La Torre, *Doing Christian Ethics from the Margins* (Maryknoll, NY: Orbis Books, 2014), 140–41.

13. George Tinker, "Spirituality, Native American Personhood, Sovereignty and Solidarity," in *Spirituality of the Third World: A Cry for Life,* ed. K. C. Abraham and B. Mbuy-Beya (Maryknoll, NY: Orbis Books, 1994), 126.

14. Clara Sue Kidwell, Homer Noley, and George E. "Tink" Tinker, *A Native American Theology* (Maryknoll, NY: Orbis Books, 2001), 33.

15. Aruna Gnanadason, "Toward a Feminist Eco-Theology for India," in *Women Healing Earth: Third World Women on Ecology, Feminism, and Religion,* ed. Rosemary Radford Ruether (Maryknoll, NY: Orbis Books, 1996), 77.

16. Kwok Pui-lan, *Introducing Asian Feminist Theology* (Cleveland: Pilgrim, 2000), 90.

17. De La Torre, *Doing Christian Ethics from the Margins,* 138–39.

18. Leonardo Boff, *Cry of the Earth, Cry of the Poor,* trans. Phillip Berryman (Maryknoll, NY: Orbis Books, 1997), xi.

19. Susan George and Fabrizio Sabelli, *Faith and Credit: The World Bank's Secular Empire* (Boulder: Westview, 1994), 98–100. It should be noted that when the memo was leaked, in an apparent attempt to minimize the damage, Summers insisted his comments were misunderstood, as he intended them to be a "sardonic counterpoint, an effort to sharpen the analysis." See Jay Johnson, Gary Pecquet, and Leon Taylor, "Potential Gains from Trade in Dirty Industries: Revisiting Lawrence Summers' Memo," *Cato Journal* 27, no. 3 (2007): 403.

20. George and Sabelli, *Faith and Credit,* 98–100.

21. De La Torre, *Doing Christian Ethics from the Margins,* 127–28.

22. Ibid., 128; Jim Yardley, "Bangladesh Pollution, Told in Colors and Smells," *New York Times,* July 15, 2013.

23. Christiana Zinner, "Water and Racialized Infrastructure in the Era of Trump," in *Faith and Resistance in the Age of Trump,* ed. Miguel A. De La Torre (Maryknoll, NY: Orbis Books, 2017), 53.

24. CERD Working Group on Health and Environmental Health, *Unequal Health Outcomes in the United States: Racial and Ethnic Disparities in Health Care Treatment and Access, the Role of Social and Environmental Determinants of Health, and the Responsibility of the State,* Report to the UN Committee on the Elimination of Racial Discrimination (New York: United Nations, 2008), 27–28.

25. Robert D. Bullard, Glenn S. Johnson, and Angel O. Torres, *Environmental Health and Racial Equity in the United States: Building Environmentally Just, Sustainable, and Livable Communities* (Washington, DC: American Public Health Association, 2011), 53.

26. Emilie M. Townes, *In a Blaze of Glory: Womanist Spirituality as Social Witness* (Nashville: Abingdon, 1995), 55.

27. Alice Walker, *Living by the Word* (San Diego: Harcourt Brace, 1981), 147.

28. George and Sabelli, *Faith and Credit,* 170.

CHAPTER TWO

Worldviews, Water, and Justice—Sustainability in a Time of Scarcity

Loring Abeyta

Mni Wiconi. It is a Lakota phrase meaning "water is life" that became popularized across the continent by Standing Rock water protectors who, in 2016 and 2017, halted the construction of the Dakota Access pipeline in North Dakota. *Mni wiconi* was not just a slogan coined on the spot to unify water protectors in their resistance to a pipeline project that would most certainly pollute and destroy the environment and sacred sites of the Lakota community at Standing Rock. In two words, *mni wiconi* conveys to Lakota people a coherent worldview, set of values, and deep cultural understanding. As Nick Estes writes, "[f]or the Oceti Sakowin, the affirmation *Mni Wiconi,* 'water is life' relates to . . . 'being a good relative.'"[1] Estes affirms that "to be a good relative is to honor the original instructions" from the ancestors.[2] What are the original instructions to humans regarding their relationship to water? Estes looks to a particular Lakota winter count for his answer.

A Lakota winter count documents significant events in the most recent passing of seasons from winter to winter. It is composed of pictures drawn on soft, tanned hide, such as buffalo or elk, and guides a community's historian in reciting the significant history of the community year by year. As Vine Deloria Jr. observed, "the story itself is important, not its precise chronological location."[3] In other words, it is not a pictorial calendar arranged in a uniform and generalized sequence. Each winter count is a unique primary document reciting the history of a particular community. It is passed from one keeper to another through generations. Each keeper, in consultation with his or her local community, decides which events of the previous four seasons are significant to add to the winter count's "framework in which a much broader history can be attached."[4] These winter counts "give a Lakota [or other national community] perspective on events at the time," and thus are a "window into Lakota culture and ways of thinking—a different way of viewing the world."[5]

The winter count named for its keeper, Battiste Good (Brown Hat), provides a window into the Lakota worldview that water is sacred (*mni wiconi*) and that water is a relative. The earliest entry in the Brown Hat winter count, dated the year 901, records a compact between the sacred *Pte Ska Win* [White Buffalo Calf Woman] and humans that "established not only the basis of customary and ceremonial laws of humans, but also how humans would exist in correct relation to the *Pte Oyate* [buffalo nation] and the nonhuman world."[6] One early picture in the winter count illustrates the centrality of *mni wiconi* to the Lakota worldview:

> Corn kernels fall from her udders into water, making the connection between humans, plants, animals, the earth, and water. Therefore, no one had the right to cede these relations to the land, water, plants, and animals—because those relations were grounded in the first compact, the first treaty with *Pte Ska Win*.[7]

Estes explains that the inclusion of water in the winter count picture is significant to Lakota people as part of their original instructions of correct relations with humans and other-than-humans. According to Estes,

> There's a constellation associated with the *Pte Oyate* or the Buffalo Nation. That constellation is actually called *tayamni*, which actually means her water, and it's a word for womb. It comes from . . . when we were made Lakota people, but it's also connected with this biological . . . notion of our relationship to water, in the sense that everyone is born in water. That's the original pictograph.[8]

Thus one can see the centrality of *mni wiconi* to the Lakota worldview. Respect and responsibility toward water as a relative is fundamental to Lakota society as well as virtually all indigenous societies. The Haudenosaunee believe that at one time in history, all creatures could communicate, and even water could communicate. Communication was the foundation of their relationship with all creatures in the world, human and nonhuman, and even the water. With relationship comes responsibility, and the necessity of seeing relationships as reciprocal instead of extractive or "zero sum."[9] As Ione Quigley summarizes this value system, "It is just so simple. It is just respecting all of the life, the universe, every drop of water, every grain of sand, and ourselves in that process."[10]

With colonization came a very different set of instructions driving Euro-Christian invaders on the continent. To these invaders, their instructions also seemed simple and coherent within their culture and worldview. There was one crucial difference between the indigenous worldview that was conveyed in the Brown Hat winter count, and the Euro-Christian worldview that ultimately dominated all social, political, economic, and religious systems in the western hemisphere (not to mention globally). This totalizing and hierarchical Euro-Christian worldview seeks converts through evangelization, both secular and religious. All must assimilate to the universalized Euro-Christian value system of anthropocentrism in order to fuel and perpetuate the structures and institutions of individualism that now reign supreme through private property and individual salvation through a Christian god. The original instructions to Lakota people, shared through the compact with *Pte Ska Win,* have been suppressed, oppressed, overwhelmed, and denigrated through this prevailing Euro-Christian worldview that seeks unlimited individual prosperity. We see the evidence of this most acutely through the broken relationship between humans and water.

According to Christopher J. Preston, environmental ethicist at the University of Montana, anthropocentrism is "an overemphasis on the centrality of human values and concerns at the expense of nonhuman ones. In ethics, anthropocentrists maintain that humans are the primary—and often the only—centers of value in the world. They deny that anything nonhuman has any value in itself."[11] Or, to put it more succinctly, editorialist Mike Rosen famously declared: "Call me human-centric if you like, but. . . the only reason to preserve the balance of nature is to sustain human life. In the absence of humans, what would it matter if earth existed?"[12] This human domination over other humans and over the resources of the earth is operationalized through methods and practices of physical extraction—extraction of resources through exploiting, harvesting, and mining both the labor of human beings and the physical elements of the earth. This human practice is diametrically opposed to the reciprocal values of an indigenous worldview and the reciprocal practices that accompany that belief system. In the anthropocentric worldview, the only rational course of action for humans is to take from nature without going through the expenditure of reciprocating something back. It is this worldview that prompted Henry Morgenthau Jr., in his inaugural remarks to the 1944 UN Monetary and Financial Conference at Bretton Woods, to proclaim that "prosperity has no fixed limits."[13]

An anthropocentric worldview has destructive impacts on the waters of the earth. In this worldview, water has no intrinsic value because it is not human. Rather, it is to be extracted and exploited for the use of humans, so that humans may prosper and flourish in isolation from the other beings and elements of the earth. Peter Brabeck Letmathe, former CEO of Nestlé, famously proclaimed:

> The one opinion, which I think is extreme, is. . .declaring water a public right. That means that as a human being you should have a right to water. That's an extreme solution. The other view says that water is a foodstuff like any other, and like any other foodstuff it should have a market value.[14]

The reduction of water from a relative to a commodity with market value is evident in cases such as the water disaster of Flint, Michigan, where water safety was sacrificed to "bottom line" financial decisions. According to Flint Rising's director, Nayyirah Shariff, the resulting water crisis left children in her community afraid of water.[15] The children of Flint, Shariff argues, have come to believe that it is bad to have water on their skin due to contaminants that burn, itch, and blister. While the physical effects of children growing up afraid of water are devastating, it is the psychological toll that is even more damaging. The children of Flint see water as an enemy, rather than as a relative.

The commodification of water has come back to haunt even the very people with whom *Pte Ska Win* made the original compact to treat all living things as relatives, including water. Beginning in 1946, with accelerating development and the continued rise of capitalist industrialization, a massive water project to deliver Missouri River water to South Dakota for agricultural, industrial, and household use was planned, funded, talked about, intermittently constructed, fought over, and ultimately never completed according to the original design. By June 1987, tribal members from the Oglala Sioux Tribe became involved, and the new phase was named the Mni Wiconi Water Project. Missouri water was to be delivered to the Pine Ridge Reservation along with private sector and other public sector customers. The result today is a minimally functioning water pipeline that supplements Missouri River water with ground water wells, especially on Pine Ridge Reservation. Racial disparities are clearly evident in the project's revival through how the Oglala Sioux Tribe has been dealt with in the current distribution of water. Seth Tupper writes that

> Although the project can deliver up to 20 million gallons of potable water daily to an estimated 52,000 people—about one-fourth of them white, and the rest Native American—it has been beset by overspending and incompletion, with the unfinished parts situated at the tribal ends of the pipeline."[16]

White communities and white ranchers receive all the pipeline water from the Missouri river, while funding constraints meant the "Oglala Sioux people were made to replace half of their share of Missouri River water with groundwater, essentially to benefit their white neighbors."[17] In the 1980s, at the time of the Mni Wiconi Water Project revival, racism played a role in the funding of the pipeline that resulted in the unequal distribution of Missouri River water. Larry Pressler, a former Republican congressman and senator from South Dakota, acknowledged that federal funding would not have been secured "if it was just going to serve whites."[18] Pressler believes that, in part, "Native Americans were used by whites to get a water pipeline approved by Congress."[19] And thus the politics of the Mni Wiconi pipeline have betrayed *Pte Ska Win's* first instructions to her human relatives.

Flint's water crisis and the mismanagement of the Mni Wiconi Water Project are just two examples of the consequences when reciprocal relationships between all beings of the earth are broken and sacrificed in order to extract value for limitless prosperity. For indigenous peoples, this colonized entrapment in an anthropocentric extractive worldview has forced them to break the first law of White Buffalo Calf Woman, that humans should "exist in correct relations to the *Pte Oyate* and the nonhuman world."[20] Onondaga Faithkeeper Oren Lyons reminds us that abandoning these relational responsibilities comes at a great cost: "From where we come from, the natural law is simple in this case: we will suffer in exact ratio to our transgressions and the damage done may be permanent to life as we know it today."[21]

How can the damage be repaired, and relationships be restored? Estes begins to answer this with questions he suggests the water protectors of Standing Rock ask us:

> What does water want from us? What does the earth want from us? *Mni wiconi*—water is life—exists outside the logic of capitalism. Whereas past revolutionary struggles have strived for the emancipation of labor from capital, we are challenged not just to imagine, but to demand the emancipation of earth from capital. For the earth to live, capitalism must die.[22]

While capitalism undoubtedly threatens the indigenous worldview that is evident in compacts such as the one between *Pte Ska Win* and the Lakota people, I argue that capitalism's demise is only part of the solution. Humans must emancipate themselves from an anthropocentric worldview if the earth is to live. The logic of reciprocity, rather than extraction, must prevail and be enacted in human actions through the rejection of individualism and hierarchy. This project transcends capitalism, communism, and all the other "isms" that ultimately place human interests above the interests of all other creatures and beings on earth. The future of the earth requires a return to the original instructions, found in indigenous experiences and cultures, that require humans to see themselves in reciprocal relationship with all elements of the earth—including water.

Scholars from diverse indigenous communities have made similar arguments, using different terminology that converges with the worldview conveyed by Brown Hat. Gustavo Esteva and Madhu Suri Prakash respond to three "sacred cows" of modernity: the myth of "global thinking," the universality of human rights, and the myth of the "individual self."[23] At the heart of their argument is the distinction between *comida* and *alimentos* as terms for food:

> The fire is at the center of the warmest room of the house. And doña Refugio is there, every day, at the very center, surrounded by her whole family. . . . And family life is at the center of the community.

> The whole community's life is in fact organized around such fires, the center of kitchens, the source of *comida*. The very essence of the *milpa* is here. . . . The essence of their "we," their *milpa*, is precisely here: around the communal fire, in the very heart of the family. . . . In contrast, *alimento* is reserved for professional, institutional, or industrial use, referring to the foods. . . eaten by the "individual self." To eat, to care for *comida*, to generate it, to cook it, to assimilate it; all these are the activities that do not belong to industrial eaters. . . . *Alimentarse*, in contrast, is to purchase and consume *alimentos* (edible objects) designed by professional "experts," while being produced and distributed through institutions for the convenience of industrial eaters.[24]

Just as the corn kernels fall from the cow's udders to the water in Brown Hat's entry, so does *milpa* mark the essence of relationship in an indigenous community a far distance south of where Brown Hat made his winter count paintings. This is a coherent and consequential indigenous worldview that holds the potential to restore relationship between humans and the other-than-human world. Without water, there is no *comida*—or no *alimentos*, for that matter. But it is the *comida* way of doing things, and following White Buffalo Calf Woman's treaty of relationship among all things of the earth, that will empower us to build a future where children are not afraid to touch and drink water.

Indeed, we can look to indigenous peoples for guidance on embodying a more reciprocal point of view toward water and its role in our lives. As demonstrated in Lakota winter counts, indigenous peoples relate to water as they do to other living beings and even humans—as kindred, rather than as "resources" for extraction and profit. The solution to water scarcity and environmental racism, then, is to vitalize such a reciprocal worldview, rather than to continue allowing the domination of a nonsustainable and destructive worldview of extraction. The latter operates out of an anthropocentrism that has shown itself to damage the world, privatizing and commodifying water to the exclusion of communities and cultures that do not share its illusions of superiority and self-sufficiency. Indeed, it is these marginalized communities that have actively resisted the extractive industries that consume, pollute, and destroy water resources. The prevailing systems that thrive by maximizing extraction, even within liberal ideologies that confront problems of racial justice, ultimately reproduce hierarchical economic, social, and political structures that place our environment in peril. To take a different course, then, we must operationalize a worldview borne of reciprocity rather than resource extraction. Indigenous nations, which demonstrate millennia of sustainable livelihoods through the wise use of water, can point a way forward.

Notes

1. Nick Estes, *Our History Is the Future: Standing Rock versus the Dakota Access Pipeline, and the Long Tradition of Indigenous Resistance* (Brooklyn: Verso, 2019), 21.

2. Nick Estes, "Standing Rock: Indigenous Futures in an Age of Apocalyptic Climate Change," Hoch Cunningham Environmental Lectures, Tufts University, March 1, 2018, https://www.youtube.com/watch?v=IXAVDkIXvSI.

3. Vine Deloria Jr., "The Concept of History," in *Spirit and Reason: The Vine Deloria, Jr., Reader,* ed. Barbara Deloria, Kristen Foehner, and Sam Scinta (Golden: Fulcrum, 1999), 290.

4. Candace Greene, in "Lakota Winter Counts," Smithsonian Education, video, 4:47, February 5, 2008, https://youtu.be/XNaYrAKiZmw.

5. Ibid.

6. Estes, *Our History Is the Future,* 83.

7. Ibid.

8. Daniel Denvir, "Our History Is the Future with Nick Estes," June 29, 2019, in *The Dig,* produced by *Jacobin* magazine, transcript and audio, 2:44:52, https://www.thedigradio.com/transcripts/transcript-our-history-is-the-future-with-nick-estes.

9. Joyce Tekahnawiiaks King, "The Value of Water and the Meaning of Water Law for the Native Americans Known as the Haudenosaunee," *Cornell Journal of Law and Policy* 16, no. 3 (Summer 2007): 452.

10. Ione Quigley, in *Oceti Sakowin: The People of the Seven Council Fires,* dir. Kyle Mork, South Dakota Public Broadcasting, 55:18, https://www.pbs.org/video/south-dakota-documentaries-oceti-sakowin-seven-council-fires.

11. Christopher J. Preston, *Grounding Knowledge: Environmental Philosophy, Epistemology, and Place* (Athens: University of Georgia Press, 2003), xi.

12. Mike Rosen, "Warming Watermelons," *Rocky Mountain News,* July 7, 2006.

13. Henry Morgenthau, "Address by the Honorable Henry Morgenthau, Jr. at the Inaugural Plenary Session," *United Nations Monetary and Financial Conference: Final Act and Related Documents,* Publication 2187 (Washington, DC: US Department of State, 1944), 4.

14. *We Feed the World,* dir. Erwin Wagenhofer (Allegro Film, 2005), 96 min.

15. Nayyirah Shariff, Conference on the Intersection of Water and Environmental Racism, Center for EcoJustice, Iliff School of Theology, Denver, CO, October 26, 2019.

16. Seth Tupper, "Unfulfilled Promise: $488 Million Water Project Incomplete for Tribes," *Rapid City Journal,* May 26, 2019, updated July 5, 2019, https://rapidcityjournal.com/news/local/unfulfilled-promise-488-million-water-project-incomplete-for-tribes/article_be19397f-d30a-5d5a-a7ee-fa86d87eaaf2.html.

17. Ibid.

18. Ibid.

19. Ibid.

20. Estes, *Our History Is the Future,* 83.

21. Oren Lyons, "Traditional Native Perspectives," *Orion Nature Quarterly* 9, no. 3 (Summer 1990): 33–34.

22. Estes, *Our History Is the Future,* 257.

23. Gustavo Esteva and Madhu Suri Prakash, *Grassroots Postmodernism: Remaking the Soil of Culture,* 3rd edition (London: Zed Books, 2014), 58.

24. Ibid., 58. *Milpa* is the indigenous, community-based system of farming in which mutually supporting plants are grown. In Guatemala, *milpa* also refers to the whole plant of maize, rather than just the kernels of corn. Both meanings of *milpa* emphasis the whole of community, whether that community is human or plant—and the connection of both.

CHAPTER THREE

Waterways—Paths of Freedom and Justice for All

Jennifer S. Leath

Many of the extraordinary properties of water are familiar to us. However, it is good to be reminded of some of them. This chapter is a meditation on some of the properties of water—and the ways that water has functioned in biological, geographical, metaphorical, and spiritual journeys toward freedom. In fact, I believe that it may be through a reconsideration of "waterways" (i.e., the *ways* of water and the ways of *water*) that it might be possible to forge new paths of freedom and maintain new equilibria of justice for all. The liberative potential of waterways is not a given, of course. In her book, *Demonic Grounds: Black Women and the Cartographies of Struggle,* Katherine McKittrick writes: "Deep space and a poetics of landscape *reposition* black geographies through taking notice of the ways in which space and place are fundamentally mentally tied to the material landscape and daily social processes" (italics original).[1] Furthermore, she continues: "The interconnections between the metaphorical and the material (for example ample real and symbolic underground[s]) bring into focus the terrain of struggle inside and outside social theory."[2] That is, the physical properties of things—in this case, water—may provide resources for reflecting upon social and moral life. Exploring the bio-logics of water is a way of engaging the "deep space" and "poetics of landscape" about which McKittrick writes. This application of this exploration dismantles the topographies of "demonic grounds." The materiality of water becomes a metaphor for liberative living. Through a remembrance of oppressive and liberating waterways, this chapter invites a renewed intentionality about cohering with and adhering to water among those committed to ecojustice as freedom and justice for all.

Cohesion and Adhesion

First, water is attracted to polar molecules, which means that water is prone to connect with other molecules.[3] This is expressed in both cohesion and adhesion—that is, water molecules are attracted to other water molecules (cohesion) and they are attracted to molecules of different substances (adhesion). This dual attraction is a consequence of water's polarity, which is the result of hydrogen bonding. Two important consequences of water's cohesiveness are surface tension—in which molecules at the surface form stronger bonds—and the liquidity of water at moderate temperatures. In addition to cohesion with other water molecules, the adhesion of water occurs when it adheres to *any* molecule with which it can form hydrogen bonds. Because of the partially positive ends of the hydrogen in water and the partially negative ends of the oxygen in most forms of glass, water has a stronger attraction to glass than it does to itself. As a result, there is a capillary action, whereby water appears to crawl up glass tubes in which it is contained. This is why we see a negative meniscus at the surface of water in a tube.[4]

It has been said—and scientists have gone to great lengths to confirm—that "opposites" attract. For water, this works in two ways. On one hand, it means that water is attracted to itself: the (slightly) positively charged hydrogen and the (slightly) negatively charged oxygen sides of water connect. On the other hand, that the positive charge of the hydrogen is stronger than the negative charge in oxygen means that water can be attracted to that which is negatively charged outside of itself. That is, water looks for difference in order to form new bonds, and it is even willing to separate from itself. As it coheres *and* adheres, water *literally* prophesies the wisdom of Mary McCleod Bethune, Mary Church Terrell, Fannie Coppin, Harriet Tubman, Ida B. Wells, and other clubwomen in and beyond the National Association of Colored Women, who coined the motto: "Lifting As We Climb." As these Black clubwomen discerned, it was essential to the best of their character and to their survival to stick together with one another and other Black women who needed support as they allowed a kind of "capillary action" to draw them up the sides of the society in which they found themselves contained.

Black feminist adrienne maree brown explains that it was *water* that invited her to complete her book, *Emergent Strategy: Shaping Change, Changing Worlds*. She writes:

> I came to this edge of Mexico to pull a book together because, a few years ago, it was here, near this tiny portion of the massive ocean, that I began to realize how important emergent strategy, *strategy for building complex patterns and systems of change through relatively small interactions,* is to me—the potential scale of transformation that could come from movements intentionally practicing this adaptive, relational way of being, on our own and with others. The waves drifting me towards and then away from the shore, seabirds of all kinds flocking over the salty blue, pelicans dropping down and scooping schools of tiny fish into their big mouths. Diving under the moving surface to see how everything humans leave in the water is repurposed as an ecosystem. Laying in a hammock and watching mosquito mamas approach me, wait for my attention to wander so they can feed on my blood (emphasis added).[5]

As "complex patterns and systems of change through relatively small interactions," brown's emergent strategy illumines the liberating *biological* possibilities of water alongside its geographically, metaphorically, and spiritually liberative capabilities. She explains that one of the core principles in learning and using the idea and method of "emergent strategy" is being like water. "Change is constant. (Be like water)," she writes, going on to cite Bruce Lee: "You must be shapeless, formless, like water. When you pour water in a cup, it becomes the cup. When you pour water in a bottle, it becomes the bottle. When you pour water in a teapot, it becomes the teapot. Water can drip and it can crash. Become like water my friend."[6] Thus, brown explains how the shape of water must shift and how water can be contained; water shifts to match its container. brown explains not only how containers of water—or anything, for that matter—change (i.e., the container that is our planet is changing), but also how *water* must change. If, as Octavia Butler posits, "God is change,"[7] it would be proper to say that water participates with God *as* change.

Heat Capacity

Second, water has an unusually high specific heat—that is, the amount of energy required to change the temperature of one gram of a substance by one degree Celsius.[8] As a result of the unusually high specific heat of water—along with its relatively high heat of vaporization—water is able to moderate temperature. It is relatively well known that boiling water takes time. This is because whatever temperature water *has* it likes to *hold*. Water tends to hold its temperature because the hydrogen bonds between water molecules either absorb or release heat—breaking or forming bonds—based on changing coherence. This process minimizes fluctuations in temperature, thus assisting in the maintenance and development of various environments and the organisms within them.[9] Though the waters can be troubled, there is a way that water attempts to maintain order, to maintain temperature, and to maintain flow. This is a frightening and exciting neutrality that characterizes water.

Of the most infamous of waterways in geographies of Blackness, McKittrick writes:

> [T]he middle passage is, obviously, not simply a theoretical concept: it is a body of water and time on a body of water, which is interconnected to black imaginative work and different forms of black politics and black travels and exiles. The meanings of the middle passage are simultaneously multiscalar and contextual. It is a geography that matters because it carries with it (and on it) all sorts of historically painful social encounters and all sorts of contemporary social negotiations.[10]

The middle passage is, perhaps, the most salient proof that waterways matter, especially as potential carriers and containers of oppression. The Atlantic Ocean carries, swallows, contains, vomits, delivers, receives, and launches the enslaved—and their stories. The middle passage and McKittrick's reflections on it remind us that geographies and their waterways have not always been kind to bodies of color. They have often been manipulated to participate in the corrosive agendas of systemically, epochally successful hegemonic *Homo sapiens*.

Beyond the specificity of the middle passage, the ideological cruelties of misunderstood and misused water must not be ignored. McKittrick explains the shock that Christopher Columbus experienced when he learned that there was land where he thought there could only be water, when he learned that there were inhabitants where he believed none could exist:

> These landmasses should have been, according to "the earth" and accepted geographies of his Latin Christian Europe, submerged:

> these areas were comprehensively non-navigable, uninhabitable, unlivable, and oceanic. Columbus's geographic "discovery"—and his personal belief that God did put land in the Western Hemisphere—radically challenged the embedded geo-religious rules of his time; geographic imaginations were stretched and forced to confront the unimaginable: landmasses above the water![11]

McKittrick continues: "The uninhabitable—the nonexistent lands and underwater places—were existent, occupied, and above water. But what becomes of the native-occupied 'uninhabitable' zones is a geo-racial reorganization."[12] And it was water—and its absence—that shaped (perhaps as much as it was shaped by) a geo-racial reorganization that exposed an embryonic white supremacy. The theological, sociological, and geographical perspectives that are brought to bear on indigenous bodies of places that are now called Africa, North America and South America, the Caribbean, Australia and New Zealand, and other iterations of "the space in between,"[13] fall like a devastating deluge on the minds of colonizers and colonized alike. In these ways that McKittrick describes, waterways are at the whim of oppressors and colonizers, used to destroy.

However, McKittrick also presents other constructed options. There are, as in the case of Linda Brent (also known as Harriet Jacobs), "garreting" possibilities. That is, there are ways that even what appear to be everyday oppressive cartographies have liberative potentials built into them. The garret, or attic, of Jacobs' free Black maternal grandmother's house became an unexpected respite. McKittrick coins the term "garreting" to describe the attic that "serves as a workable paradoxical space, one that positions [Jacobs] across the regime of slavery and begins her emancipation. . . . [T]he garret is a painful geographic expression of black emancipation."[14] McKittrick argues that through garreting "the geographic perspective of slavery is widened and understood as an alterable geographic device—a new geographic form—which counters the naturalization of race, gender, and place."[15]

Water, too, because of its high specific heat can function as a garreting device—and, I would argue, it *is* just the kind of garret that is necessary for so many on the Underground Railroad. In more than one of her biographies, the story is told that Harriet Tubman, born Araminta Ross in 1822, was tasked with wading in water to cut muskrats. Kate Larson, one of her biographers explains:

> Trapping muskrats would be a difficult task for a young child, but in cold water during the winter trapping season, when muskrat pelts are at their finest, it would be even more so. At some point Tubman became sick with measles but was forced to continue working the traps in the cold water. She became gravely ill. Rit [Harriet's mother] then convinced Brodess to bring little Harriet home.[16]

Catherine Clinton describes the same circumstance. However, it is also clear that Tubman learned the water. She *had* to learn the water. Clinton explains that Tubman "confided that she had observed that all the streams she knew ran north to south."[17] This knowledge of water taught her how to escape running against the stream. On one of her thirteen-plus missions to free seventy or more enslaved people, Larson recounts the story:

> [S]he always seemed to know when danger was near; her heart would "go flutter, flutter." Some thirty miles from Wilmington, probably in the vicinity of Blackbird, her intuition led her to immediately change her planned route. "God told her to stop," which she did, "but then he told her to leave the road, and turn to the left." As they approached a "small stream of tide water," they looked for a boat or a bridge to cross over but found none. Their only option was to wade across, but it was cold and the depth of the stream was unknown to them, so the men refused to go into the water. Tubman, convinced they had no choice and that the risk of capture was too great, plunged in and waded across without them. The water nearly reached her shoulders, but no higher, and she safely reached the other side. Seeing that she had not drowned, the men reluctantly followed.[18]

On one of her most impressive missions, leading Black troops in the Union Army, Tubman suffered devastating numbers of casualties at the Fort Wagner battle in 1863, but she was also able to orchestrate the escape of 700 enslaved people through her extraordinary intelligence, expressed in skills such as espionage, strategy, memorization, geography, and waterway navigation.[19] Of those like Tubman, pressed to follow the pull of water toward freedom—the specific heat of water notwithstanding—Vincent Harding writes: "The old voices in the Atlantic's waves were again inviting, compelling responses from their kindred on the ship."[20] It is as though spirits of the bodies drowned, thrown overboard dead, jumped in to their death boiled the ocean in anger past that high specific heat, jostling the waters to the defense of the oppressed, enslaved, and colonized. Harding describes Tubman's journey:

> In 1849 she broke free. . . . Wading through waters, she made her way to Philadelphia. . . . As Harriet Tubman joined the Northern and Southern movements in her own scarred body, the God of the river appeared to have many voices—perhaps one for each eager listener—and many tasks in the struggle to free a people from their bondage.[21]

The tasks meant mastering the cartographies, geographies, and waterways known to slavers and dogs even better than the slavers and dogs did. The tasks meant keeping a spiritual flame on the waterways that would frustrate any social neutrality.

Heat of Vaporization

Third, water has a relatively high heat of vaporization. This is the other quality, besides water's high specific heat, that enables it to moderate temperature. We might think of the heat of vaporization as the amount of heat energy needed to change a gram of liquid into gas.[22] For water, this is well known to occur in a boil at one hundred degrees Celsius. Just as water needs a lot of energy to change temperatures by even one degree Celsius up or down, the specific change of water's form from liquid to vapor (or gas)—which is the result of a breakdown of water's hydrogen bonds—requires a great deal of energy. Moreover, the process of evaporation produces a corresponding cooling effect on the surface of whatever the previously liquid water had been in contact with. Water not only tends to stay the same temperature, but when it evaporates it causes a cooling—another form of moderation with respect to temperature.

When the heat passes out of water through the process of vaporization, a cooler, more moderate substance is left behind. In the face of oppression, a healthy and appropriate response is anger—anger that expresses itself in movement, anger that expresses itself in change, anger that expresses itself in heat, and (hopefully more often than not) anger that expresses itself in love. The heat of vaporization means that some of what was water in solid or liquid form no longer remains; what has vaporized dissipates into earth's atmosphere. Evaporated water becomes a part of the breathed ecosystem of the air; sometimes evaporated water condenses on surfaces and oozes out of spaces we may not have expected; sometimes evaporated water, without the right ventilation, condenses and can give way to the soured scents of mold, mildew, and other trapped particles. To be free of solid and liquid water forms, regardless, requires a great deal of energy.

We might consider such energy like that which would be required to separate a mother from her child—or a child from her mother—at the border. Water, most stable as ice or liquid, is dispersed as gas. It cannot take the heat; it must not take the heat. And though there *is* connection preserved in the vaporized form, in this earth's season of unusually inappropriate warming, we must be doing all we can to keep water together in its liquid and solid forms. We must be doing all that we can to discourage the angry separation of water as a result of evaporating heat while using the high specific heat of water, the moderating effects of water, to channel "fleeing humanity" and living creation along paths of freedom and justice. If and when water *does* evaporate, we must not deny the cool, the chill, the positively liberating changes enabled through such transition—even if the heat of anger was released with evaporation.

A traditional negro spiritual dating back to slavery contends:

Sometimes I feel like a motherless child
Sometimes I feel like a motherless child
Sometimes I feel like a motherless child

The final line of this spiritual varies. As early as 1899, William Barton's score reads: "Den I gi' down on my knees and pray."[23] More familiar to the modern ear, Paul Robeson sings: "A long ways from home, a long ways from home."[24] In her book *Lose Your Mother* (2008), Saidiya Hartman repeats the same story of a woman lost in the middle passage three times. She compels the reader to read it, to read it, and to read it again. Why? Because this woman is a lost mother, a lost daughter, and there is something impossible and demonic about this.[25]

So serious is this relationship of mother to child that Miguel A. De La Torre includes the following story about Yemayá in his book, *Santeria: The Beliefs and Rituals of a Growing Religion in America.* I offer the story as an extended quote to clarify just how serious the evaporation of relationship between mother and child can be and how—at great lengths—it has been maintained as water to itself and water to the watered.

> One of the most well-known legends about Yemayá concerns her adopted son Changó. As a young boy, Changó ran away from his parents Obatalá and Yemmu. He was discovered by Yemayá, who took him in and reared him as her own son. In time he became a man and went out into the world to seek his fortune. He traversed the earth, and as he did so he had numerous amorous relationships. Becoming an incurable womanizer, he began to spend most of his time at parties, dancing and seducing women. On the rare occasions when he wasn't making love, he was out making war. Many of the orishas surmised that his actions were the result of his anger toward his birth mother Yemmu, who he felt had abandoned and betrayed him all those years ago.
>
> One day he was at a party when he noticed Yemayá. He was smitten then with her mature beauty. In an instant he forgot her years as his stepmother and became intent on seducing her. Yemayá was not impressed, and knowing about the life of heavy drinking, brawling, and skirt chasing he had been living, she decided to teach him a lesson. She suggested that they leave the party and go to a place where they could be alone. Delighted at such an easy conquest, Changó followed Yemayá.
>
> She led him to a boat on the edge of a lake, telling him that her home was on its far shore. Hesitantly Changó, who disliked water, climbed aboard. When the boat reached the middle of the ocean, Yemayá plunged into the sea and disappeared below the waves. Changó grew frightened, for he did not know how to swim or how to handle a boat. And then he noticed a gigantic wave approaching. The wall of water crashed down on the tiny vessel. Knocked off the craft, Changó found himself sinking to the ocean floor. Desperately he paddled to the surface and managed to pull himself back into the boat. It was then that he saw Yemayá approaching him, riding yet another wave. Changó beseeched her to help him. "I will help

> you," she replied, "under one condition: that you learn to respect your mother." He cried out, "My mother? But she abandoned and betrayed me when I was just a boy!" Then Yemayá reminded Changó that he had two mothers, one who birthed him and one who raised him. Then Changó recognized Yemayá and repented for the way he had treated her—and for the way he had been treating women in general. The anger and bitterness he held toward his birth mother Yemmu dissipated. To this day he loves to seduce women, but because of Yemayá he shows much more respect for them than before.[26]

De La Torre not only explains the importance of this mother-child relationship, but also just how important Yemayá is to understanding the earth after an initial beginning that ended in flames of destruction. He writes:

> Where the fires burned most fiercely, the grounds were left with lower elevation. As the rains fell, these low points filled with water, creating the great oceans we see today. Here, in the huge crevices between the cooled rocks, the orisha Yemayá was born. Yemayá laid her body across the earth and exclaimed, "My womb is hurting!" Then from her womb came forth all the rivers, orishas, and creatures great and small.[27]

The water as motherly is inextinguishable. He writes:

> Water cleanses. It washes away dirt, as well as evil energy, cleansing both the physical and the spiritual. Fresh water, the domain of the orisha Oshún, becomes the source of fertility. Salt water, on the other hand, under the rule of the orisha Yemayá, symbolizes the maternal, the source of all life. Stagnant water, such as that in a lagoon, represents spiritual death—which is a prerequisite for new life. And finally, the earth, the fourth element, provides the necessary resources for survival.[28]

So it is no surprise that Araminta Ross, affectionately known as "Minty," likely had a religious conversion of sorts and, soon after she was married, took up her mother's name: Harriet, affectionately known as "Rit," a name that had been carried throughout their family, the name of the one that defended the young Tubman to the master who had her wading for muskrats until she almost died.

In his book, *The Water Dancer,* it is as though Ta-Nehisi Coates has heard and tapped into these two stories—of Yemayá with Changó and of Araminta turned Harriet Tubman. Enslaved people called "the Tasked" call Hiram, the protagonist, Moses. Hiram has conduction power which makes the earth fold "like fabric" when the "conduction" takes place.[29] He transports himself and other people across great distances when he taps into the power of *remembering* his mother. The evaporation of the relationship between mother and child, between water liquid and water vapor is, at times and in certain ways, unnatural. For freedom and justice, in some instances, relationship must be restored.

Ice Density

Fourth, ice, the solid form of water, has a relatively low density—that is, relative to liquid water. Below 0 degrees Celsius, water's hydrogen bonds form ice crystals. Due to the stability of those bonds at that temperature, those crystalline structures are easily maintained, despite the increased distance between those bonds. As a result, ice floats on water; its solid form is less dense than its liquid form.[30] Thus icebergs float, the top of lakes freeze, falling into chilled water under ice can kill, and ice-fishing can thrill.

Water is more than itself. While it is most often thought and spoken of in terms of its liquid form, its solid form is equally important. And when it is not solid throughout, the float of solid water atop liquid water can be dangerously deceptive. It can appear that surface ice is stable to we who would skate across it, even when this stable glass is thin and shallow. When broken and regathered, its crystalized edges can be sharp and jagged, devastatingly cutting into that which it contacts. This is possible because ice is not as heavy (in the sense of density) as water. Though dangerous when shallow or shattered, in its earliest stage, perhaps ice performs one of the miracles of Jesus—walking on water.

Water is more than a liquid, more than a life indicating substance. Consider the geographic danger of the oceans and seas as middle passage for people of African descent; consider the ways water became a cloak of erasure for indigenous people fleeing from and drowned in the arrival of genocidal colonizers. Water can be friend or foe to those seeking and maintaining freedom. Through her book, *The Black Shoals: Offshore Formations of Black and Native Studies*, Tiffany Lethabo King explores shoals as a metaphor to connect the plight of enslaved people of African descent and indigenous people from this "Turtle Island" of whom but a remnant have survived in the face of horrific and persistent genocide. King opens with the following definition of "shoal": "*n.* A place where the water is of little depth; a shallow, a sand-bank or bar. *adj.* of water, etc.: not deep; shallow. *ex.* The boat could not come to land the water was so shoale."[31] King explains:

> Before the invention of bathymetry, the calculation of probability curves that estimated the latitude and longitude of shoals—rock formations, coral beds, and sandbars—were very difficult to map and, therefore, avoid. For captains, sailors, and others on board a ship, the perils of the sea included crashing into a shoal and the sinking of their vessel. . . .
>
> The word "shoal" has a number of meanings. Geological sources define it as the area in which the sea or a body of water becomes shallow. As a process, it is the movement of the ocean from greater to shallower depths. It is not the shore; it is a formation before the shore or offshore. As a location and geological formation it is often described as a sandbar or a coral reef. It is an accumulation of granular materials (sand, rock, and other) that through sedimentation create a bar or barrier that is difficult to pass and, in fact, a "danger to navigation." As a geological unit, it is a physical place, a shallower place in the ocean before one arrives at the shore. . . .
>
> While also used to describe nongeological matter such as a school of fish, the term is rarely used in humanistic terms, however. Declining in use after the eighteenth century, the word "shoal" is generally used in the form of a verb to describe how a ship or vessel slows down to navigate a rocky or rough seabed that has risen toward the surface of the ocean. As the waters became shallow, a ship would shoal to avoid running aground.[32]

The shoal mirrors from the base the ice that sits atop liquid water's surface. When shallow or shattered, the shoal can be devastating for those who encounter it. Though the shoals are denser than liquid and solid water, they still demonstrate the floating that is possible in water, the ways that water visibly conceals its solid or liquid forms when they are together, and the ways that an equilibrium is possible when solid and liquid are understood together instead of separately. As King might put it, justice for one must mean justice for all, especially when one is of African descent or indigenous and all are indigenous or of African descent. Again, as King might argue, it is inadequate to float the metaphors of water for people of African descent and land for indigenous people; the shoals clarify that we are essentially, fundamentally, and necessarily together. Water is a fundamentally hybrid and potentially dangerous substance—even when the hybridity extends beyond the simultaneous presence of water as solid, liquid, and gas; it can be shoals, shallow and full of diverse texture—water not water; it can be Black and indigenous at the same time.

Polarity

Fifth, water itself has a relatively high polarity. Water responds to polar molecules, where there is a difference of electronegativity between the atoms, and ionic compounds, where there are atoms carrying negative and positive charges.[33] Water responds to full charge, partial charge, and polar molecules.[34] The polarity of water makes it an ideal solvent. Sodium chloride (NaCl), commonly known as salt, is a great example of an ionic compound that water dissolves. When exposed to water, water's hydrogen attaches to, separates, and surrounds the sodium and chloride. To us, it simply tastes like salt water.

When water is attracted to polar molecules, it is attracted to that which is strangely like itself. The hybridity of polarity is not just reserved for the oppressive other. The hybridity of polarity is found in the colonized and oppressed other. It is hard to discern if the hybridity of polarity in colonized and oppressed others is a direct consequence of colonization and oppression—expressing what might be described as good and evil, a unique alterity gifted with opaque seventh sight only because of the sufferings endured, and/or that positive and negative charge just *always* express themselves in *everything*. Whatever the case may be, water has polarity. It expresses itself in positive and negative charge; it expresses itself in a unique alterity bound together as two hydrogens and an oxygen. And its polarity—in the presence of other polarities—responds to the material and charges with which it comes into contact. In the case of salt, on one hand, water dissolves it; on the other hand, the flavor of salt infects the water with which it mixes. Water that has encountered and mixed with salt, settling in solid or liquid form and temperature is merely diluted salt. The two substances—salt and water—at temperatures below boiling and/or below circumstances suitable for evaporation, are inseparably bound; they are together. Such inseparable togetherness in the midst of difference, alterity, hybridity, and/or opacity is the strangeness that Homi Bhaba and Charles Long explore.[35]

W. E. B. Du Bois, late nineteenth- and early twentieth-century sociologist and Black theoethicist of sorts appreciated the complexities of water in its strange hybridity. Du Bois tells of a Black man, Jim Davis, who survives a comet crash and is left as the apparent lone survivor with a white woman in the short story "The Comet." Jim survives because he has been sent into a watery vault when the comet is about to hit—while his employers prepare to watch the comet with delight. The dialogue goes:

> "Oh, that was Halley's," said the president; "this is a new comet, quite a stranger, they say—wonderful, wonderful! I saw it last night. Oh, by the way, Jim," turning again to the messenger, "I want you to go down into the lower vaults today. . . . Everything of value has been moved out since the water began to seep in," said the president; "but we miss two volumes of old records. Suppose you nose around down there—it isn't very pleasant, I suppose."[36]

As the story goes, when Jim is alone with the only other apparent survivor, he reaches a point of resignation:

> Yet as the two, flying and alone, looked upon the horror of the world, slowly, gradually, the sense of all-enveloping death deserted them. . . .
>
> Great, dark coils of wire came up from the earth and down from the sun and entered this low lair of witchery. The gathered lightnings of the world centered here, binding with beams of light the ends of the earth. The doors gaped on the gloom within. He paused on the threshold.
>
> "Do you know the code?" she asked.
>
> "I know the call for help—we used it formerly at the bank."

> She hardly heard. She heard the lapping of the waters far below,—the dark and restless waters—the cold and luring waters, as they called. . . . He pointed down to the waters, and said quietly:
>
> "The world lies beneath the waters now—may I go?"
>
> She looked into his stricken, tired face, and a great pity surged within her heart. She answered in a voice clear and calm, "No."[37]

These two, like opposites, are attached—like polar water with salty ions. This reluctant attachment continues until the white woman's father emerges—and Jim's wife later. Then there is an evaporation that reminds us that hope is fleeting at best. And the ever-present mediator for this familiar story of hybridity is water: the water that conceals, that traps, and that promises freedom. The polarity of water holds opposites together, even when they no longer desire to be bound; the polarity of water invites the separation of those bound together, even when they are unethically connected.

Conclusion

Within dark water there is also a sound vibration that has invited the creative authorship of those who have moaned and groaned at the door of freedom. Of spiritual waters and the living water of spirituals, W. E. B. Du Bois writes: "Then there is the song of many waters, 'Roll, Jordan, roll,' a mighty chorus with minor cadences."[38] At the heart of *The Souls of Black Folks* is a meditation on such sorrow songs: songs of grief over family, songs of commitment to freedom. Many of these are songs of waterways. Clinton notes that songs were a signal familiar to Tubman's freedom journeys.[39] One song Tubman may have sung when it was time for slaves running to move was "Wade in the Water." A song that Barbara Holmes cites as a freedom song for the enslaved and a connection to water, the sky, swimming, and flying is "Follow the Drinking Gourd."[40] In its own ways, each of these songs makes sense of the natural properties of water for the natural pursuits of freedom and justice.

And consider those whom adrienne maree brown cites as she follows the lead of waterways to freedom and justice. She quotes Malkia Cyril: "From water I have learned to move around and past fixed objects to reach my goal."[41] brown recalls the words of Desiree Evans:

> All these streams and rivers making their way back to the source, to the goal, to the collective vision of the sea. In building our movements, we learn this: we come from different places across the country and world, we face different obstacles, paths, twists, and turns—some of us reshaping the very rocks beneath us, some of us moving the rocks with our combined strength, and some of us going around them when they can't be moved and finding another path to our desired goal.[42]

And she, herself, recognizes the complexity of the stuff that makes up two thirds of our bodies: "Remember you are water. Of course you leave salt trails. Of course you are crying. Flow."[43] Then there is this quote of Corina Fadel: "Resilience: (v) The way the water knows just how to flow, not force itself around a river rock; then surely I can stretch myself in the shape my own path is asking of me."[44] The song lyrics and melodies—along with other words penned for the freedom and justice that water portends—are remnants of ecomemory, invitations to retrieve our ecomemories.

According to Melanie Harris, we have opportunities to retrieve our ecomemories and rediscover a common moral vision. Especially drawing on the experiences of people and communities of African descent, Harris writes:

> Spirit, nature, and humanity are connected in an interdependent web of life in African cosmology. Thus, any ethical or unethical behavior conducted by humans impacts the other aspects of the cosmological order positively or negatively. According to this framework, one could argue that since ancestors are believed to reside in many aspects of nature, any human behavior that diminishes and dishonors nature or the earth can have a devastating impact on the relationship between the human and the ancestor. In the case of water pollution, for example, the act of humans misusing, damaging, wasting, or abusing water is understood to be an immoral act against nature, which disrespects the ancestors. In this sense, the common ethical worldview that Peter J. Paris describes has woven within it an ethical mandate to care for the earth. This important ethical frame for earth justice also can be said to be innate to African and African American life; thus, an ethical mandate to care for the earth is considered normative in many African cultures and communities.[45]

While this ethical call to care may engender hope, we do well to remember brown's water wisdom once more: "that water seeks scale, that even your tears seek the recognition of community."[46] Caring for those who are crying, mourning, and suffering is no easy task. Still waters run deep, deep calls out to deep, water seeks the sea, and water seeks its level and equilibrium. A lot transpires as these missions of water run their course. Harris quotes Mercy Oduyoye:

> In my tradition when there is a flood, a river overflows, we always say the river is angry, the river is angry and when you look you will find that they have blocked the river someplace with rubbish that people have thrown in or something. So that's what I mean by, if nature is nice and beautiful—I feel ok. . . . I feel good when I'm in a place where nature looks good and is happy.[47]

How many of us can claim such happiness in the waterways of our lives? Perhaps, recognizing these principles of water—(1) that water is attracted to polar molecules; (2) that water has an unusually high specific heat; (3) that water has a relatively high heat of vaporization; (4) that ice, the solid form of water, has a relatively low density—that is, relative to liquid water; and (5) that water, itself, has a relatively high polarity—we might make waterways of freedom and justice for all.

Notes

1. Katherine McKittrick, *Demonic Grounds: Black Women and the Cartographies of Struggle,* 1st ed. (Minneapolis: University of Minnesota Press, 2006), 17.

2. Ibid., 18

3. Lisa A. Urry et al., *Campbell Biology,* 11th ed. (New York: Pearson, 2016), 45. Though this and the other properties of water emphasized in this essay are verified and developed in my study of *Campbell Biology,* I wish to acknowledge that these properties of water were also brought to my attention (as a set) through an informal blog: livingconsciously, "5 Properties of Water," *Owlcation* (February 7, 2019): https://owlcation.com/stem/5-Properties-of-Water.

4. "Capillary Action and Why We See a Meniscus," *Khan Academy,* https://www.khanacademy.org/science/biology/water-acids-and-bases/cohesion-and-adhesion/v/capillary-action-and-why-we-see-a-meniscus.

5. adrienne maree brown, *Emergent Strategy: Shaping Change, Changing Worlds* (Chico, CA: AK Press, 2017), 2.

6. Ibid., 41, citing Bruce Lee, *Bruce Lee: A Warrior's Journey* (Warner Home Video, 2000).

7. Octavia Butler, *Parable of the Sower* (New York: Grand Central, 2007 [1993]), 3.

8. Urry et al., *Campbell Biology*, 46.

9. Ibid.

10. McKittrick, *Demonic Grounds*, 18.

11. Ibid., 128.

12. Ibid., 129.

13. Ibid., 46–47.

14. Ibid., 144.

15. Ibid.

16. Kate Clifford Larson, *Bound for the Promised Land: Harriet Tubman, Portrait of an American Hero* (New York: Random House, 2004), 38.

17. Catherine Clinton, *Harriet Tubman: The Road to Freedom* (New York: Little, Brown, 2004), 37.

18. Larson, *Bound for the Promised Land*, 128–29.

19. Clinton, *Harriet Tubman*, 176–80.

20. Vincent Harding, *There Is a River: The Black Struggle for Freedom in America, Reissue edition* (Boston: Mariner Books, 1993), 153.

21. Ibid.

22. Urry et al., *Campbell Biology*, 47.

23. William Barton, *Old Plantation Hymns: A Collection of Hitherto Unpublished Melodies of the Slave and the Freedman, with Historical and Descriptive Notes* (Boston: Lamson, Wolffe, 1899), 17.

24. Paul Robeson, "Sometimes I Feel Like a Motherless Child," arr. Lawrence Brown, Victor 20013-A, 1926, 78 rpm.

25. Saidiya Hartman, *Lose Your Mother: A Journey along the Atlantic Slave Route* (New York: Farrar, Straus and Giroux, 2008).

26. Miguel A. De La Torre, *Santería: The Beliefs and Rituals of a Growing Religion in America* (Grand Rapids, MI: Eerdmans, 2004), 72–73.

27. Ibid., 36.

28. Ibid., 5–6.

29. Ta-Nehisi Coates, *The Water Dancer* (New York: One World, 2019), 19, 276.

30. Urry et al., *Campbell Biology*, 47–48.

31. Tiffany Lethabo King, *The Black Shoals: Offshore Formations of Black and Native Studies* (Durham, NC: Duke University Press, 2019), 1.

32. Ibid., 1–3.

33. "Molecules and Compounds," Khan Academy, accessed January 14, 2020, https://www.khanacademy.org/science/chemistry/atomic-structure-and-properties/introduction-to-compounds/a/paul-article-2.

34. Urry et al., *Campbell Biology*, 45, 49.

35. Homi Bhabha, *The Location of Culture* (New York: Routledge, 1994); Charles Long, *Significations: Signs, Symbols, and Images in the Interpretation of Religion*, Philosophical and Cultural Studies in Religion (Aurora, CO: The Davies Group, 1999).

36. W. E. B. Du Bois, *Darkwater: Voices from Within the Veil* (New York: Harcourt, Brace & Howe, 1920), 253–54.

37. Ibid., 266–67.

38. W. E. B. Du Bois, *The Soul of Black Folks,* ed. Henry Louis Gates Jr. (New York: Oxford University Press, 2007), 124.

39. Clinton, *Harriet Tubman,* 89.

40. Barbara A. Holmes, *Race and the Cosmos: An Invitation to View the World Differently,* 1st ed. (Harrisburg, PA: Bloomsbury T&T Clark, 2002), 87–89.

41. brown, *Emergent Strategy,* 68.

42. Ibid., 269–70.

43. Ibid., 109.

44. Ibid., 123.

45. Melanie L. Harris, *Ecowomanism: African American Women and Earth-Honoring Faiths* (Maryknoll, NY: Orbis Books, 2017), 14–15.

46. brown, *Emergent Strategy,* 110.

47. Harris, *Ecowomanism,* 104.

CHAPTER FOUR

Water Flows East to West—Interdependence

Garam Han

Water is the source of life. Water is an indispensable element for surviving, just like air. According to a NASA fact sheet, about 70 percent of the surface of our planet is water, just as the human body comprises about 70 percent water.[1] If someone loses 12 percent of their body's water content, they may lose their life.[2] Water helps our bodies to undergo metabolic changes by eliminating toxins. Metabolism works as water moves from the mouth, to the stomach, then to the heart, passes to the blood, is carried by the cells, is brought back to the kidneys, and finally is discharged. Literally, the water we drink goes through all the organs and cleanses our bodies.[3]

This information about water is well known because we hold this knowledge intuitively, so people may think it is unnecessary for it to be mentioned. However, we are living in a world where we must remind people of the importance of water, specifically *clean* water. There are many communities that have had their physical and mental body, family body, community body, and nature body washed with *toxic water*. One example is that of Flint, Michigan, where a water crisis began in 2014 and continues today. According to the *New York Times*:

> Flint's water crisis, which started in 2014, was a failure of government at all levels. A state-appointed emergency manager switched the city's drinking water source from Detroit's municipal water system to the Flint River in an effort to save money. Local officials failed to implement corrosion controls, allowing lead to leach from pipes. Health agencies assured residents the water was safe even as people complained that it smelled bad, tasted funny and was discolored."[4]

After a test conducted by the Environmental Protection Agency and Virginia Tech in 2015, they announced there are "dangerous levels of lead" in the water.[5] The lead water moved from the mouth, to the stomach, and then to the heart, where it passed to the blood, was carried by the cells, brought back to kidneys, and remained in the physical and mental bodies of the residents of Flint. The residents are poisoned by both toxic water *and* toxic power within the social system. Toxic water not only affects humanity but also the rest of creation. There are already natural sources of lead, but humans add more lead to nature from careless governing as witnessed during the Flint Water Crisis. This interdependent relationship between humanity and the rest of creation shows that toxic water hurts everyone and everything involved.

Again, we are instinctively aware of the importance of water. Yet how did the people who decided to change the water source for Flint make such a reckless decision? We use water to cook, wash, drink, and plant in our daily lives. What made the officials decide on change so easily and try to solve the problem so carelessly? One obvious answer is because they do not live in Flint, Michigan. Because they do not think Flint is their community.

The voices of Flint residents were not important to them until it became a major issue nationally and people started to blame their governmental system. Even though the officials started to reinvestigate the water, Flint residents did not have an opportunity to participate in change, and their voices were not involved from the beginning. Flint residents have already lost trust in the state government because their voices were not heard. Michigan started to do actual work on the problem only once pressure built from beyond the Flint community. Even though they addressed the water crisis, the State of Michigan experienced a failure in governing. People started to point out the fundamental problem behind the careless governing decision: environmental racism.[6] According to the US Census Bureau, Flint's residents in 2018 were mainly nonwhite (Black: 53.9 percent; two or more races: 4.3 percent; Hispanic or Latinx 3.9 percent) and "disproportionately poor."[7] When we acknowledge these statistics concerning Flint, we cannot ignore the question: If Flint were a city of mostly white and rich people, would the officials have made the same decision? Whenever the powerful face negative economic issues, minorities become the target sacrificed to protect the financial benefits of the privileged. Money was more important than people, as capitalism threatened minority groups. The Flint Water Crisis proved the State of Michigan tried to recover from their financial crisis by *poisoning* the minority group; however, in the end, the state had to spend even more to provide bottled water, refund water bills, and cover the residential expenses.[8]

Although officials removed the lead from the pipeline, it would be hard to remedy the wounds and doubts in people's hearts. To prevent this type of catastrophe in the future, we must act against the fundamental problem of an anthropocentric and individualist mindset that attacks the whole community. In this chapter, I suggest that a different perspective—one that is nature-centered and communal rather than anthropocentric and individualist—may be observed in Korean philosophies and religions—namely, shamanism, Donghak, and *salim*. In so doing, I will rely on Kim-Cragg's work concerning interdependence as a foil over against the prominent Western (anthropocentric and individualist) perspective.

Korean Shamanism

Shamanism, as the indigenous religion of Korea, existed before the tenth century BCE. The shamanistic ritual is an event for whole communities.[9] During the ritual, the shaman becomes a mediator and communicates with the spirit world.[10] This shamanistic ritual is still practiced in Korea. Many people visit shamans for good energy when they begin a new journey, such as a new job, starting a family, or the start of the new year. On the other hand, people also seek shamans when their job or family has problems.

A Korean theologian, Choi Hee An, understands the image of shamanistic gods as a protectors and benefactors:

> Shamanistic gods are strongly associated with the roles of benefactor and protector. People believe their gods protect them from bad fortune, bad weather, harmful ghosts, evil countries, and so on. Thus, they become not only protectors from evil but also bring about worldly blessings and foster social harmony. . . . Because of continual poverty and the threats of colonialism, the Korean people traditionally have had a strong need for both spiritual and psychological

> protection. This explains why all the shamanistic gods are closely connected with every aspect of the people's daily lives: they provide both protection and stability.[11]

Korean shamanistic rituals are mostly regarded as a community event, since Korean indigenous people care not only for the personal but also for the communal healing and well-being.[12] Therefore, the protector and benefactor gods are *communal* protectors and benefactors, not personal gods.

The shamanistic ritual is also represented in domestic settings, and one of the rituals was *junghwasu* 정화수 (a bowl of clean water). Women in a Korean family prepared the very first water drawn from the communal well in the early morning in preparation for prayer. The reason they prepared the very first water of the day was that they wanted to worship gods with *clean* water, before anyone used it for personal purposes. The Korean indigenous people regarded water as a mediator between gods, representing the perspective that nature was more than just a ritual element. *Clean* water was holy.

If the water in the communal well were contaminated, they could not worship the gods. To worship gods with polluted water is an act of unfaithfulness.[13] Therefore, taking care of nature and the environment becomes an unquestionable duty for the Korean indigenous people. However, if the communal well was polluted by a natural catastrophe, they believed the contaminated water meant evil things had happened, and they tried to find what injustice happened in the community, attempting to fix the problem together. This cultural understanding about the relationship with environment and nature is naturally embodied in Koreans. As mentioned above, the Korean indigenous people believe in many gods, and one of the images of gods is the god in nature—like gods of the rivers, mountains, and heavens. Since Korea was for centuries an agrarian culture, those natural images of gods were close to the common people, such as farmers and the working class, throughout their everyday lives.[14]

Furthermore, *junghwasu* on a *jangdokdae* (Korean earthen jar) or in the kitchen is the common image most Korean people imagine in the *junghwasu* prayer.[15] *Jangdokdae* was used for containing sauces and kimchi in Korea. I interpret this as the way Korean indigenous people worship gods within their common life. They did not need a temple or building to pray. The indigenous religion was a part of their everyday lives. Another important characteristic is that *junghwasu* prayer is not only for their family but also for the community. Praying for the community is more than a shamanistic characteristic. It is a cultural philosophy and ideology that reflects a communal perspective, which will be further explored below.

As we see, Korean shamanism possesses a nature-centered and a communal worldview. Because of the various images of gods, being humble before nature and people is a virtue in Korean culture. Therefore, dominating the world, harming and destroying nature and other beings were regarded as actions against gods. Korean indigenous religion, as well as traditional Korean art, presents nature as a larger figure compared to humanity. Some works of art show humanity as smaller than the animals or only depict nature in the picture.

However, Korea's communal and nature-centered perspective has been subjugated by "Western Protestant Christian practices (particular ways to worship, pray and sing, for instance)" as Protestant missionaries from the West came to Korea.[16] Protestant Christians in Korea condemned shamanistic rituals as evil, idolatrous practices and strove to eradicate them.[17] The biggest reason for this opposition is that shamanism has multiple gods, which contrasts with Christian monotheism. Although missionaries tried to contaminate Korean shamanism, this communal and nature-centered perspective was deeply embedded in Korean culture, as revealed in different religions and ideologies. Donghak and *salim* are good examples.

Donghak

Donghak, which literally means "eastern learning," was founded by Jae Woo Choi in 1860 and is called "Chondogyo" today. Choi started a national religion, Donghak, over and against the colonization and westernization by Japan and Christianity in the nineteenth century.[18] Jae Sophia Oh claims Donghak is "inter-religious, inter-cultural, inter-ideological, and intertraditional" because Choi brought the understandings of Asian philosophical religions and Christianity from the West, then Koreanized those understandings to form Donghak.[19]

Donghak teaches that *hanulim* 하늘님 (the divine) is present within each one of the "human beings, animals, plants, and even non-organic bodies."[20] Finding God in all beings teaches Koreans to value the equal dignity of all creatures. Choi's understanding of the divine is "not as a static ontological being but as one who keeps working in an incomplete sense, therefore, becoming." Choi also emphasizes daily life rather than life after death, so Donghak believers tried to live their lives fully within a "life-centered and biocentric view."[21] Donghak believers contributed to Korean history by actively leading and participating in political movements for justice such as the Donghak Peasant Revolution (1894), which was a movement for human rights by a minority group, and the Korean independence movement against Japan.[22] Donghak's fundamental idea that it is right to serve each person equally and serving *hanulim* led believers see the injustice in society.[23]

One of the essential rituals in Donghak is the *junghwasu* prayer from Korean shamanism. Believers in Donghak prepared a bowl of clean water for an offering to *hanulim*. Choi's idea about clean water in *junghwasu* has three main themes: the communal, individual, and healing.[24] First, Choi wanted to change Korean social structure with the *junghwasu* faith.[25] Neo-Confucianism was the main religious philosophy in Korea during the nineteenth century, and it has a patriarchal and hierarchal ideology.[26] In the Neo-Confucian ritual, women prepared the elements needed for worship, but they could not participate in the worship service because of their gender. Therefore, Choi made a space for women to participate and contribute to worship by including the *junghwasu* in the ritual, since a bowl of clean water for praying in a shamanistic ritual was prepared through women's own initiative. Although Korean women in Donghak were not perfectly freed from the patriarchal and hierarchical ideology, having a space for women to participate in worship was a major step for that time.[27]

Another social challenge in Donghak by way of the *junghwasu* faith was to make service simple to reduce its financial burden.[28] Ancestor worship in Neo-Confucianism placed the financial and physical burden of acquiring food and fruits for the ritual on the women and the poor, making it difficult for some to participate. On the other hand, the simple way of worshipping the divine with a bowl of clean water was a way to lessen a load for the people. Compared to Neo-Confucian worship, the worship in Donghak was easy to prepare, requiring only clean water (at that time) from the communal well, thus eliminating the financial burden.

Secondly, Choi taught people to build good relationships among individual family members by sharing the bowl of clean water they used for worship.[29] Donghak believers conducted service every day in their houses, just like *junghwasu* prayer in Korean shamanism. However, unlike the *junghwasu* in shamanism, Donghak devotees shared clean water to bring intimacy, respect, faith, and trust to their family. Donghak focuses on the mutual relationship with the divine (*hanulim*) through a relationship with others.

Thirdly, Donghak followers believed the bowl of clean water after worship could heal the people mentally and physically. At the time Choi initiated Donghak, most Koreans were very poor and mentally and physically sick from war and colonization. Therefore, the belief in healing with a bowl of clean water after worship was attractive to the people. Choi really believed in the power of the clean water as a curative means, and many miracles were reported. People were cured by a bowl of clean water used for worship. As people witnessed these miracles, *junghwasu* became a more symbolic element in Donghak.[30]

Donghak was a Korean religion for the people, especially for the minority groups in society. They did not fear to challenge the powerful and injustices within societal systems. Donghak's doctrines emphasize the communal and nature-centered ideas among not only humanity but also the rest of the creation. Sophia Oh claims Donghak has an "ecocentric and cosmocentered" philosophy, and it influenced a new understanding of *salim*.[31]

Salim

Salim 살림 is a Korean word, a noun version of *salida* 살리다, which means "save, rescue, salvage; to safeguard from damage or injury."[32] So *salim* means "making things alive, restoring and enlivening."[33] The word *salim* has been devalued and defined as women's household chores, like cooking, cleaning, and raising children. However, many Korean and Korean American scholars—like Ji Ha Kim, Hyun Kyung Chung, HeeSung Chung, Jae Sophia Oh, and Hee Sun Kim[34]—are reclaiming the word "*salim*" as a positive way to recognize its ability and power to make things alive outside of the patriarchal and hierarchical system. *Salim* is working in daily life and doing indispensable works.

Transforming the meaning of *salim* was initiated by Ji-Ha Kim, a Korean poet and activist who was against the dictatorship in the 1970s and 1980s.[35] The poet Kim was influenced by the ideology of Donghak, which values everyday life in the present rather than life after death. Kim started to see the divine in people's daily lives. Donghak's "eco-centric and cosmocentered philosophy" enabled Kim to see the "degendered and devalued" meaning of *salim* from a different perspective when he was in a prison because of his involvement in the human-rights movement.[36] The poet Kim realized death is a *part* of life, not the *opposite* of life. *Jugim,* which means "intentional killing and murder," is the opposite of life. Kim adopted the idea of *jugim* (killing life) in the concept of *salim* (restoring life). Oh claims that "*Jugim* is not natural but exists contrary to nature and the unnatural activity of life engages violence. *Salim* represents all diverse activities of life which include intentional efforts to overcome *jugim* and thus restore life and all its spontaneous changes and natural growth."[37] If we stopped making an intentional effort to overcome intentional killing, only *jugim* would be left, and the violence from *jugim* would be repeated in our society.

We must do *salim* and have to invite people to do *salim* together to prevent the violence from *jugim.* The importance of "intentional effort" to do *salim* is that it should be enacted in our daily lives.[38] *Salim* is very practical for our lives, just like eating, drinking, and washing. However, what is there to prevent our life of *salim* today? What is the system that prevents our basic and practical life today? What is the system of injustice in our society that prevents the basic rights of people, specifically the minority? Who and what is it in the system of *jugim,* intentional killing?

Interdependence

The common idea among shamanism, Donghak, and *salim* is to have an *interdependent* worldview as their roots. Hyeran Kim-Cragg, in her book *Interdependence,* argues that "interdependence of life" is an important way to become a healthy community by challenging independence. Kim-Cragg believes humans are interdependent with each other because of the "the limit of human existence."[39] She emphasizes that humanity must recognize its limits and that, furthermore, those limits are not negative. Rather, such limits fully acknowledge our interdependence. Kim-Cragg introduces a Korean anthropological perspective about humanity's interdependent relationships through a Korean-Chinese etymology of a person, 인간 (人間).

> A person "인간," "人間" (pronounced as *InGan* in Korean) consists of two words, "人" (in) meaning "a person" and "間" (gan) means "between." The character "人," by depicting two lines leaning on one another, suggests that we are persons by virtue of the fact that we lean on each other. It may be interpreted to mean people cannot stand alone—we need another to rely on. When we put the two characters, "人," (in) and "間" (gan) together we get "person-between" suggesting what makes us human are the relationships between us.[40]

A Korean-Chinese etymology of a person, 인간 (人間), represents well the Korean understanding of humanity, which is based on a communal idea. A person cannot exist without others. I exist because you exist. We exist because a community exists. A community exists because the other community exists. Human community exists because a community of the rest of creation exists. Kim-Cragg argues that the full personhood can be formed through the interdependent relationship, so we must "live out a restoration of the community."[41]

Kim-Cragg also suggests the idea of interdependence beyond an anthropocentric perspective. She emphasizes recognizing the community of creation in the interdependent relationship by criticizing human's ignorance and violence, which are *jugim*, toward the rest of creation community. Unfortunately, the ignorance and violence mostly affect marginalized groups of human and the rest of creation, and the marginalized have started to lose their healthy communities because they cannot live the fullness of life. Their dignity is destroyed by the environmental injustice under the western system of anthropocentricism, capitalism, individualism, and dualism.[42] Kim-Cragg strongly states that "anthropocentrism is a fundamental problem of ecological crisis and harming human-earth-divine relationships."[43] Anthropocentrism brings a hierarchical view, which prevents interdependent relationships among creatures and builds a wall between different beings, ethnicities, classes, religions, and genders. Destroying our relationship with creation also destroys humanity's capacity to be who they are. As a result, they cannot see the interdependent relationship even among their fellow humans.

Moreover, we must have a macro and micro perspective for understanding interdependence in order not to be poisoned by an "anthropocentric bias."[44] I did not learn of the Flint Water Crisis until several years after it became a major news story in the United States. I ignored such issues in the United States with the excuse that I am an alien in this country. Comparatively, I checked Korean news every day while living in the United States as an international student. Other issues, however, like President Trump's desire to build a border wall and lack of concern for gun violence and the climate crisis, have garnered my attention. However, I did not learn about the Flint Water Crisis in the discussion of justice issues in my theological setting. Perhaps I missed the conversation, but this also means that the conversation was not repeated enough for me to remember in my community. There are many global issues we must discuss, of course, but it is telling that we are so busy talking only about the issues that directly affect us and our own communities because we are living with an "anthropocentric bias" and our interdependent, communal relationships are limited to a micro perspective.

Like the idea of Korean shamanism, Donghak, and *salim,* we should re-root our worldview in a larger, macro perspective and acknowledge the interdependence of our relationships. In the Flint Water Crisis, officials have worked out this issue to resort to their own abilities and skills because they did not see the interdependent relationship they shared with the Flint community. To better reflect this interdependent relationship, the residents should have a safe and clean space to raise their voices and to engage in the changes. This means officials should stop thinking about new methods coming from their heads but listen to the residents' voices and opinions, and perhaps they could even learn to care about those in communities other than their own.

Notes

1. NASA Fact Sheet, "Follow the Water: Finding a Perfect Match for Life," National Aeronautics and Space Administration (April 16, 2007): https://www.nasa.gov/vision/earth/everydaylife/jamestown-water-fs.html.

2. SaeChang Oh, "What Is Water for Human?" *Gyeongju Society* 6 (July 2018): 68.

3. Ibid.

4. Mitch Smith, "Flint Water Prosecutors Drop Criminal Charges, with Plans to Keep Investigating," *New York Times,* June 13, 2019.

5. CNN Editorial Research, "Flint Water Crisis Fast Facts," *CNN,* updated August 23, 2020, https://www.cnn.com/2016/03/04/us/flint-water-crisis-fast-facts/index.html.

6. Charles Blow, "The Poisoning of Flint's Water," *New York Times,* January 21, 2016.

7. "QuickFacts: Flint city, Michigan," US Census Bureau, https://www.census.gov/quickfacts/flintcitymichigan.

8. CNN Editorial Research, "Flint Water Crisis Fast Facts."

9. Su Yon Pak, Unzu Lee, Jung Ha Kim, and Myung Ji Cho, *Singing the Lord's Song in a New Land: Korean American Practices of Faith* (Louisville: Westminster John Knox Press, 2005), 71.

10. Christine J. Hong, *Identity, Youth, and Gender in the Korean American Church: Asian Christianity in the Diaspora* (New York: Palgrave Pivot, 2015), 65.

11. Hee An Choi, *Korean Women and God: Experiencing God in a Multi-Religious Colonial Context* (Maryknoll, NY: Orbis Books, 2005), 15.

12. Pak, Lee, Kim, and Cho, *Singing the Lord's Song,* 71.

13. Pil-young Lee, "Symbolism of Well in Korean Folklore," *Korean Historical Folklore Institute* 26 (March 2008): 253–54.

14. Choi, *Korean Women and God,* 14.

15. Ibid., 17.

16. Pak, Lee, Kim, and Cho, *Singing the Lord's Song,* 71.

17. Ibid.

18. Jea Sophia Oh, *A Postcolonial Theology of Life: Planetarity East and West* (Upland, CA: Sopher, 2011), 2.

19. Ibid.

20. Ibid., 3.

21. Ibid., 24.

22. James N. Poling and HeeSun Kim, *Korean Resources for Pastoral Theology: Dance of Han, Jeong, and Salim* (Eugene, OR: Pickwick, 2012), 47–48.

23. Ibid., 47.

24. Eung-Tae Jo, "The 'Water Faith' of Korean New Religions," *The Korean Academy of New Religions* 22 (April 2010): 157.

25. Ibid.

26. Neo-Confucianism is a departure from the earlier Confucian belief, becoming more hierarchal and patriarchal. Neo-Confucianism teaches that "the subordination of female to male was taken as 'natural.'" As NeoConfucianism was introduced into Korean culture and religion via China during the Joseon Dynasty (1392–1910), "the distinction of social-class" was observed, and women's rights and activities became extremely limited. According to a Korean theologian Ryu Tong Shik, during the late Joseon Dynasty, Neo-Confucianism presented Koreans with meaning and direction for life. The influence of Neo-Confucianism became a life ethic in Korean culture and continued to suppress women. See Greer Anne Wenh-In Ng, "Asian Sociocultural Values: Oppressive and Liberating Aspects from a Woman's Perspective," *People on the Way: Asian North Americans Discovering Christ, Culture, and Community,* ed. David Ng (Valley Forge, PA: Judson, 1996), 68–69; Tong Shik Ryu, *The Christian Faith Encounters the Religions of Korea* (Seoul: Christian Literature Society of Korea, 2001), 88.

27. Jo, "'Water Faith,'" 157.

28. Ibid.

29. Ibid., 158.

30. Ibid., 34.

31. Oh, *Postcolonial Theology of Life,* 18.

32. Poling and HeeSun Kim, *Korean Resources for Pastoral Theology,* 75.

33. Oh, *Postcolonial Theology of Life,* 2.

34. There are several others; here I only list scholars mentioned in this chapter.

35. Oh, *Postcolonial Theology of Life,* 18.

36. Ibid., 18–19.

37. Ibid., 19–20.

38. Ibid., 21.

39. HyeRan Kim-Cragg, *Interdependence: A Postcolonial Feminist Practical Theology* (Eugene, OR: Pickwick, 2018), 28.

40. Ibid., 28–29.

41. Ibid., 29.

42. Ibid., 131.

43. Ibid., 133.

44. Ibid., 143.

WATER AS A HUMAN RIGHT

CHAPTER FIVE

The Right to Water under International Law and Environmental Racism

Ved P. Nanda

Introduction

Inequalities and discrimination are prevalent not only in developing countries or in the United States but on every continent, demanding deep reflection on the manner in which water rights are impacted by environmental racism everywhere. After a review of the current water crisis around the globe and the water situation we face here in Colorado—regarding both water scarcity and poor water quality—this essay will discuss international action on water, including the right to water under international law. I will then present a critical look at the situation in this country and how environmental racism deprives people of color and migrants, non-English speakers, and the poor and the marginalized of the benefits of social justice, including access to affordable clean water. The conclusion suggests that civil society can indeed muster an effective response to the challenges over water that the world currently faces.

Water Scarcity and Poor Water Quality the World Over

Water is life—it is the most basic of human needs. The United Nations Water program (unwater.org) has aptly stated that "water is a precondition for human existence and for the sustainability of the planet."[1] It reports that 700 million people worldwide could be displaced from their homes by intense water scarcity by 2030.[2]

The global water crisis is a crisis of water scarcity and poor water quality. Water scarcity results from limited water resources that are inadequate to meet the demand. It could also result from inadequate water infrastructure or poor management of water resources. Many countries and several rapidly growing megacities all over the world are facing water shortages caused by population growth, pollution, and unpredictable and extreme weather patterns caused by climate change.[3] In the last couple of years, Chennai in India, Basra in Iraq, and Cape Town in South Africa have suffered severe water crises. Without urgent action it is feared that "Day Zero" could become a reality in many parts of the world. *Scientific American* describes that milestone as the "date when the city's taps would go dry because its reservoirs would become dangerously low on water."[4] Official reports are that by 2020, New Delhi, Bangalore, and Hyderabad, among twenty-one cities in India, will run out of groundwater.[5] Drought and poor infrastructure maintenance is creating a national water crisis in South Africa.[6]

Several organizations studying water scarcity have sounded alarms. To illustrate, according to the *United Nations World Water Development Report 2019*, four billion people—that is, two-thirds of the global population—live under conditions of severe water scarcity at least one month in a year.[7] The report finds that in 2015 more than two billion people worldwide lacked access to safely managed drinking water services, and 844 million lacked access even to a basic drinking water service.[8] What a travesty that 263 million people spend more than 30 minutes walking to collect water from an improved source and 159 million still collect drinking water directly from surface water sources, according to a 2017 World Health Organization study![9]

The World Economic Forum in its Insight Report, *The Global Risks Report 2019*, lists the water crisis as one of the ten top risks in terms of likelihood and top fourth risk in terms of its impact.[10] Based on a review and analysis of what the World Business Council for Sustainable Development (WBCSD) calls "a wide range of publications by respected institutions," the WBCSD identifies water one among five emerging "hot sustainability topics" and thus "areas of potentially increasing pressure for companies."[11] It projects that by 2030, "global demand for fresh water [will] exceed supply by 40 percent," that "drought affected many places in 2018," that "[r]ising sea levels are impacting coastal cities more regularly," and that, in addition to Cape Town, ten other global cities are already at risk of "water day zero."[12]

The World Resources Institute reported in 2019, based upon new data, that seventeen countries, constituting one-quarter of the world's population, face *extremely* high water stress, while forty-four countries, home to one-third of humanity, face high levels of stress.[13] Twelve of those seventeen countries are in North Africa and the Middle East, and climate change is likely to further complicate matters for them.[14] With the existing climate change scenario, one-third of the world's largest groundwater systems are already in distress.[15]

Andrew Liveris, chief executive of Dow, a chemical company, has called water "the oil of the twenty-first century."[16] Furthermore, a recent issue of *The Economist* featured a special report suggesting that disputes over water will be an increasing source of international tension.[17] It notes that Ariel Sharon, the commander of the Six-Day War of 1967 and subsequently Israel's prime minister, wrote in his autobiography that the war "really started on the day Israel decided to act against the diversion of the Jordan. . . . The matter of water diversion was a stark issue of life and death."[18]

The Economist cites studies by the Pacific Institute, a think tank in Oakland, California, discussing the use of water as a weapon by a terrorist group that diverted water from the Jubba River in Somalia, causing a flood that forced opposing forces to move to higher ground, where they were ambushed.[19] It also notes that conflicts over water can lead to violence, as has happened in Kenya and Nigeria, and that transboundary tensions and conflicts can occur within a nation, for example, as in the Nile, Ganges-Brahmaputra, Indus, Tigris-Euphrates, and even in Colorado.[20]

Conservation Colorado recently released a report card on Colorado's eight rivers, warning that they are threatened by overuse, climate change, poor dam management, energy development, and the growing population, which is expected to double by 2050.[21] The Colorado River received a "D" grade.[22] Adding to the challenge are Colorado's water policies of "first come, first served," and "use it or lose it."

The *Wilson Quarterly*, however, provides an encouraging outlook:

> Beyond the frightening headlines of "water wars," real work is being done to anticipate and mitigate future water conflicts, both between and within nations. The Wilson Center is collaborating with NOAA and the University Corporation for Atmospheric Research to develop a framework to improve predictive capabilities for security risks posed by extreme weather events, including drought and flooding.[23]

In addition to water scarcity, poor water quality aggravates the problem. The World Bank reports that the impacts of poor water quality are "wider, deeper, and more uncertain than previously thought and require urgent attention."[24] Pollution can cause GDP in downstream regions to be reduced by one-third to half, threatening growth.[25] Salinity reduces agricultural productivity, imperiling food security. Considering that 80 percent of all wastewater is not treated before it returns into the environment, it harms public health, especially in developing countries, where water-related diseases are widespread.[26]

The awareness of the water crisis has challenged governments to find solutions and take necessary initiatives but, for developing countries, these challenges are immense. Consider, for example, that they lack adequate information; sufficient human and financial resources and sound fiscal management; appropriate policies, strategies, and plans of action; and adequate legislation, regulations, and effective measures. Many countries suffer from inadequate capacity, corruption, lack of coordination and planning, and—equally important—inappropriate technology choices.

Clearly, a critical need is to address the root causes of water insecurity: pollution, overuse of water, travel distance to water sources, drought, and floods, among others. Political solutions include groundwater recharge, ensuring aquifer storage and recovery, water conservation, rainwater harvesting and water catchment, recycling wastewater, properly managing ecosystems, and, of course, population growth control and climate change mitigation. Among other solutions is desalination, but it is extremely costly, which developing countries can ill afford, and water privatization, which neither ensures more access to water nor results in economic growth in developing countries.[27] The technological challenges are indeed formidable, but even if they can be surmounted, it is the political side of the equation that will decide whether the promise of clean water for all becomes a reality. It is indeed easier said than done.

International Action and the Right to Water under International Law

International action on water includes the 1977 United Nations Water Conference in Mar del Plata, Argentina, which proclaimed: "All peoples, whatever their stage of development and their social and economic conditions, have the right to have access to drinking water in quantities and of a quality equal to their basic needs."[28] Since that time, several conventions, treaty monitoring bodies, and resolutions have recognized the right to water.

To illustrate, the 1979 Convention on the Elimination of All Forms of Discrimination Against Women ensured the right of water supply to women as it called upon states parties to "take all appropriate measures to eliminate discrimination against women in rural areas in order to ensure, on a basis of equality of men and women, that they participate in and benefit from rural development."[29] Ten years later, in November 1989, the Convention on the Rights of the Child explicitly stated that states parties shall pursue full implementation of the right to water.[30] Next, in January 1992, the International Conference on Water and the Environment in Dublin, Ireland, adopted the Dublin Statement on Water and the Environment, stating among its four guiding principles that "it is vital to recognize first the basic rights of all human beings to have access to clean water and sanitation at an affordable price."[31] Seven years later, in 1999, the United Nations General Assembly affirmed that "[t]he rights to food and clean water are fundamental human rights and their promotion constitutes a moral imperative both for national Governments and for the international community."[32] The next year, the Sub-Commission on the Promotion and Protection of Human Rights recognized the right to drinking water and sanitation.[33]

The UN General Assembly proclaimed the period from 2005 to 2015 an International Decade for Action called "Water for Life." Following this, the General Assembly declared another International Decade for Action beginning in 2018, "Water for Sustainable Development," emphasizing that water is critical for "the eradication of poverty and hunger, that water, energy, food security and nutrition are linked, and that water is indispensable for human development, health and well-being."[34]

In July 2010, the UN General Assembly formally recognized the human right to water as indispensable for leading a life in human dignity and acknowledged that clean drinking water is a prerequisite for the realization of all human rights.[35] No country voted against the resolution as it passed by 122 voting in favor and 41, including the United States, abstaining. This was followed by the Human Rights Council's action in September 2010 affirming the decision of the General Assembly and confirming that the human right to water and sanitation is binding as part of existing international human rights law.[36] The council called upon states to "achieve progressively the full realization of human rights obligations related to access to safe drinking water and sanitation."[37] The right to water requires that physically accessible and affordable water that is sufficient and safe be available for both personal and domestic use.

Among recent noteworthy developments, the General Assembly set the Millennium Development Goals for the period from 2000 to 2015, aimed at halving the proportion of the population without access to water and sanitation by 2015.[38] The outcome was that 147 countries met the drinking water target, 95 countries met the sanitation target, and 77 countries met both, and thus the rate of global population using an improved drinking water source rose from 76 to 91 percent.[39] As a successor to the Millennium Development Goals, the Sustainable Development Goals—a set of seventeen goals and 169 targets addressing social, economic, and environmental aspects of development—were adopted by UN member states on September 25, 2015, as part of a comprehensive and ambitious package, the 2030 *Agenda for Sustainable Development*.[40] The sixth of these goals, "Ensure Availability and Sustainable Management of Water and Sanitation for All," is a standalone water goal designed to achieve by 2030 "universal and equitable access to safe and affordable drinking water for all,"[41] and improve water quality.[42] Specific indicators for monitoring the progress and measuring the outcomes are now in place.[43]

In 2019, UN Secretary-General Antonio Guterres reported that progress in the proportion of safely managed drinking water services increased from 61 to 71 percent between 2000 and 2015, but the number of people still lacking safe water, sanitation, and hand-washing facilities remains in the billions.[44] This still leaves 785 million people who lack even a basic drinking water service; the proportion lacking even a basic sanitation service still stands at 27 percent; and those lacking basic hand-washing facilities at home are estimated at three billion people.[45] The Secretary-General reports that one-third of countries have medium or high levels of water stress, all of them located in North Africa, West Asia, or in Central and South Asia.[46]

Introducing the Secretary-General's report at the High-Level Political Forum, Liu Zhenmin, a UN Under-Secretary-General, underscored the reality that "people living in fragile States are twice as likely to lack basic sanitation and about four times as likely to lack basic drinking-water services as people in non-fragile situations."[47] And fragile states are the ones where people of color are the overwhelming majority.

In August 2019, the theme for the prestigious annual World Water Week in Stockholm was "Water for Society: Including All." With 4,000 participants from 138 countries, it highlighted the importance of inclusion, calling it the key to addressing today's challenges.[48] There, the expression "nothing about us without us" was frequently used to underline how ethnic minorities, people with disabilities, and marginalized groups should be viewed as experts on their own needs when it comes to water and sanitation.[49] There was growing concern at the conference that climate change hurts poor and marginalized groups disproportionately and that, without action, we can expect widening gaps between countries and regions and within countries.[50] Focusing on economic, environmental, and social issues, the conference addressed the question, "How can water play a pivotal role in addressing inequalities, discriminatory practices and unjust distributions of power to ensure implementation of the 2030 Agenda?"[51] It also wrestled with another critical question, "How does water contribute to achieving equal power relations that, by including everybody, ensures fairness and breaks inequality barriers preventing human rights and equality for all?"[52] The preceding discussion shows significant international attention to the right to water as a human right. There is a global consensus that concerted action is essential to accomplish the goal of safe and affordable drinking water for all and this must be achieved soon.

Impact of Environmental Racism on Access to Affordable Clean Water

Millions of Americans are exposed to unsafe drinking water, which can cause serious and costly health issues. And several studies have found that minority and poor communities across the United States are disproportionately affected by polluted water. *Watered Down Justice*, a 2019 report by a partnership of the Natural Resources Defense Council, Environmental Justice Health Alliance, and Coming Clean, found race to be a primary factor in enforcement of the Safe Drinking Water Act.[53] This act, passed by Congress in 1974, was designed to protect public drinking water across the United States, but as the report found, the law clearly had not been enforced equally for everyone. *Watered Down Justice* analyzed nationwide violations of the act from 2016 to 2019:

> Our analysis also revealed that of all the sociodemographic characteristics analyzed, racial, ethnic, and language vulnerability had the strongest relationship to slow and inadequate enforcement. . . . Racial, ethnic, and language vulnerability . . . bore the strongest relationship to average number of quarters with violations, number of violations, and average number of rules in violation.[54]

The report divided these violations into three categories—health-based, monitoring and reporting, and public notification.[55] At the county level, for people of color, low-income people, and non-native English speakers, the rate of drinking water violations increased. The report found that "in addition to high rates of violations, the relationship between racial, ethnic, and language vulnerability and persistence of violations was the strongest of all the relationships . . . studied across all compliance, enforcement, and socio-demographic characteristics."[56]

After conducting an extended analysis of Flint, Michigan, a poor community with 40 percent of the city's population living below the poverty line, and where twelve people died because of contaminated water and 6,000 to 12,000 children were exposed to lead in their water, the report called the majority-Black (56 percent) city's deadly lead water crisis a "microcosm of the state of our drinking water system."[57] It also studied Newark, New Jersey, another predominantly Black city, which has also been facing a lead water crisis,[58] and Sandbranch, Texas, a historically Black town near Dallas that has never had public water lines or sewage service.[59] The report's conclusion is telling:

> For communities already facing severe burdens due to racism, social conditions and/or environmental and health hazards, the inability to turn on a tap and receive clean, safe water is particularly devastating—and unjust. Our findings are consistent with the long-standing pattern of disproportionate and cumulative hazards in communities of color and low-income communities. It is therefore incumbent upon our policy makers, water utilities and other responsible parties to protect the most vulnerable populations and ensure their right to safe water.[60]

Environmental justice research has shown that communities of color and low-income communities face disproportionate environmental burdens that range from contaminated groundwater to high concentrations of hazardous facilities.[61] In response to water scarcity and lack of access, many corporations have begun to privatize water systems. The Public Citizen Group gives ten reasons to oppose water privatization: 1) it leads to rate increases; 2) it undermines water quality; 3) companies are accountable to shareholders, not consumers; 4) it fosters corruption; 5) it reduces local control and public rights; 6) private financing costs more than government financing; 7) it leads to job losses; 8) it is difficult to reverse; 9) it can leave the poor with no access to clean water; and 10) it would open the door to bulk water exports.[62]

However, recent reports are that people around the world are reclaiming their access to water as a basic human right as they are seizing control over their water from private companies. Between 2000 and 2015, in 235 cases, a city, region, or national government terminated or refused to renew water concessions, leases, or management contracts with private companies, thus bringing water back under public control.[63] This covered 100 million people across thirty-seven countries now benefiting from water as a public good rather than a private commodity.[64] According to another recent study, based upon empirical evidence on privatization in developing countries, private ownership alone will no longer automatically generate economic gains in developing economies, and the literature reflects a more cautious and nuanced evaluation of privatization.[65] Yet another study finds that more privatization of water provision in developing countries will not ensure more access to water.[66]

Clearly, numerous studies have shown that race is *the* most important factor in failures to provide adequate and healthy water supplies around the world. And while the debate continues on the pros and cons of privatizing water distribution, privatization generally exacerbates the problem.

Conclusion

The planet is suffering from a crisis of water scarcity and poor water quality. People of color and poor and marginalized people bear the brunt. Notwithstanding municipal, national, and international action to address the challenge, a large percentage of the population among and within nations is still deprived of access to affordable and safe water. Without concerted and expedited action, the goal to ensure availability and sustainable management of water and sanitation for all by the year 2030 will not be achieved. It is imperative, therefore, that civil society and human rights activists work together to create awareness and appreciation of the devastating impact of environmental racism, which deprives vulnerable and marginalized people of their inherent right of access to safe and affordable water. Decision makers must be compelled to ensure that on water issues, social justice is the guidepost.

Notes

1. Julian Cribb, *Food or War* (New York: Cambridge University Press, 2019), 63.

2. Elizabeth Hameeteman, *Future Water (In)Security: Facts, Figures, and Predictions* (Columbus, OH: Global Water Institute, 2013), 3.

3. In 2019, Colorado had a record-setting high temperature in September and a record setting low temperature in October. Chris Bianchi, "It's Official: Colorado Has a New Record Hottest Temperature," *Denver Post,* October 15, 2019; Sam Tabachnik, "Pretty Unprecedented: Metro Residents Contend with October Weather for the Ages," *Denver Post,* October 30, 2019.

4. "Cities Need to Prepare for Water Day Zero, Cape Town's Recent Water Crisis Highlights the Need for Better Urban Planning and Management," *Scientific American* 319, no. 2 (August 1, 2018), https://www.scientificamerican.com/article/cities-need-to-prepare-for-water-day-zero.

5. See, e.g., Al Jazeera, "India Is Running Out of Water, Fast: Delayed Monsoon Rains Highlight Fragility of the Water Supply in the Deeply Agrarian South Asian Country," June 20, 2019, https://www.aljazeera.com/news/2019/06/india-running-water-fast-190620085139572.html.

6. Luke Daniel, "South Africa's Looming Water Crisis: Drought and Dire Infrastructure," *The South African,* October 28, 2019.

7. Richard Connor, Stefan Uhlenbrook, and Engin Koncagül, *Leaving No One Behind: The United Nations World Water Development Report 2019, Executive Summary* (Perugia, Italy: UNESCO World Water Assessment Programme, 2019), 2.

8. Ibid., 2, 4.

9. Nada Osseiran, "2.1 Billion People Lack Safe Drinking Water at Home, More Than Twice as Many Lack Safe Sanitation," World Health Organization, *World Health Organization News Release,* July 12, 2017. The news release is referencing the 2017 Joint Monitoring Programme (JMP) report.

10. World Economic Forum, *The Global Risks Report 2019,* 14th ed. (Geneva: World Economic Forum, 2019), 5.

11. Julian Hill-Landolt, Daniel Fule, and Maria Mendiluce, *2019 Outlook & Trends* (Geneva: World Business Council for Sustainable Development, 2019), 2, 12.

12. Ibid., 13.

13. Rutger Willem Hofste, Paul Reig, and Leah Schleifer, "17 Countries, Home to One-Quarter of the World's Population, Face Extremely High Water Stress," *World Resources Institute,* August 6, 2019.

14. Ibid.

15. "With One-Third of Largest Aquifers Highly Stressed, It's Time to Explore and Assess the Planet's Groundwater," *National Geographic Society Newsroom,* July 9, 2015.

16. Anonymous, "Business: Running Dry; Business and Water" *The Economist* 388, no. 8594 (August 23, 2008): 53.

17. Anonymous, "Water Fights" *The Economist* 430, no. 9132 (March 2, 2019): S8.

18. Ibid.

19. Ibid.

20. Ibid.

21. Conservation Colorado, *Colorado's Rivers: A Report Card* (2017), 2, http://conservationco.org/wp-content/uploads/2019/06/Rivers_English2018.pdf.

22. Ibid.

23. Lauren Risi, "Beyond Water Wars," *Wilson Quarterly* 43, no. 3 (Summer 2019): 1.

24. International Bank for Reconstruction & Development–World Bank (IBRD), "Quality Unknown: The Invisible Water Crises," *US Fed News Services,* August 20, 2019, www.worldbank.org/en/news/feature/2019/08/20/quality-unknown.

25. Ibid.

26. Connor, Uhlenbrook, and Koncagül, *Leaving No One Behind,* 2.

27. Stockholm International Water Institute (SIWI), *World Water Week 2019: Water for Society, Including All—Overarching Conclusions* (Johanneshov: Stockholm International Water Institute, 2019), 10.

28. United Nations Water Conference. *Report of the United Nations Water Conference, Mar del Plata, 14–25 March 1977,* Sales No. E.77.II.A.12 (Geneva, CH: United Nations Publication, 1977), http://www.ielrc.org/content/e7701.pdf.

29. General Assembly Resolution 34/180, UN Convention on the Elimination of All Forms of Discrimination against Women, Article 14(2)(h), https://www.ohchr.org/en/professionalinterest/pages/cedaw.aspx.

30. Convention on the Rights of the Child, November 20, 1989, 1577 UNTS, 3, https://www.ohchr.org/en/professionalinterest/pages/crc.aspx.

31. International Conference on Water and the Environment, the Dublin Statement on Water and the Environment, Principle 4, UN Doc. A/Conf. 151 PC /112 (January 31, 1992), http://www.wmo.int/pages/prog/hwrp/documents/english/icwedece.html#p4.

32. GA Res. 54/175, UN Doc. A/RES/54/175 (December 17, 1999), https:/undocs.org/A/RES/54/175.

33. Sub-Commission on Human Rights, Rep. of the Sub-Commission on the Promotion and Protection of Human Rights on Its Fifty-Second Session, at 31, UN Doc. E/CN.4/Sub.2/2000/46 (Nov. 23, 2000), http://undocs.org/en/E/CN.4/Sub.c2/2000/46.

34. GA Res. 71/222 (February 7, 2017).

35. GA Res. 64/292 (July 28, 2010).

36. Human Rights Council Resolution 15/9, UN Doc. A/HRC/RES/15/9 (October 6, 2010).

37. Ibid.

38. GA Res. 55/2, UN Millennium Declaration (September 8, 2000), https://undocs.org/A/RES/55/2.

39. UN Development Programme, *The Millennium Development Goals Report 2015,* 7, https://www.un.org/millenniumgoals/2015_MDG_Report/pdf/MDG%202015%20rev%20(July%201).pdf.

40. United Nations General Assembly Resolution 70/1, "Transforming Our World: the 2030 Agenda for Sustainable Development," UN Doc. A/RES/70/1 (September 25, 2015), https://www.un.org/ga/search/view_doc.asp?symbol=A/RES/70/1&Lang=E [hereafter 2030 Agenda].

41. Ibid. at Goal 6.1.

42. Ibid. at Goal 6.3.

43. Sustainable Development Goal 6, Sustainable Development Knowledge Platform, https://sustainabledevelopment.un.org/post2015/transformingourworld.

44. United Nations, Report of the Secretary-General on SDG Progress 2019, Special Edition 14–15 (2019), https://sustainabledevelopment.un.org/content/documents/24978Report_of_the_SG_on_SDG_Progress_2019.pdf.

45. Ibid.

46. Ibid.

47. UN DESA, Introduction of Secretary-General progress report on the SDGs High-Level Political Forum (July 9, 2019), https://www.un.org/development/desa/statements/.

48. SIWI, *World Water Week 2019,* 10.

49. Ibid.

50. Ibid., 11.

51. Ibid., 18.

52. Ibid., 20.

53. *Safe Drinking Water Act* (1974): https://legcounsel.house.gov/Comps/Safe%20Drinking%20Water%20Act-(Title%20Xiv%20Of%20Public%20Health%20Service%20Act).pdf.

54. Kristi Fedinick, Steven Taylor, and Michele Roberts, *Watered Down Justice* (New York: National Resources Defense Council, 2019), 55.

55. Ibid., 25.

56. Ibid., 7.

57. Ibid., 9.

58. Ibid., 29.

59. Ibid., 26.

60. Ibid., 35.

61. Amy Vanderwarker, "Water and Environmental Justice," in *A Twenty-First Century US Water Policy,* ed. Juliet Christian-Smith and Peter H. Gleick (New York: Oxford University Press, 2012), 52, 55.

62. Public Citizen, *Top Ten Reasons to Oppose Water Privatization,* 1–7, https://www.citizen.org/wp-content/uploads/Top_10_PDF.pdf.

63. Rapid Transition Alliance, *Turning the Tide of Water Privatization—The Rise of the New Municipal Movement* (January 28, 2019), https://www.rapidtransition.org/stories/turning-the-tide-of-water-privatization-the-rise-of-the-new-municipal-movement/.

64. Ibid.

65. Saul Estrine and Adeline Pelletier, "Privatization in Developing Countries: What Are the Lessons of Recent Experience?" *The World Bank Research Observer* 33, no. 1 (February 2018): 73.

66. Sylvia Cesar, "Privatization of Water: Evaluating Its Performance in the Developing World," *Annals of Public and Cooperative Economics* 90, no. 1 (July 2018): 6–7.

CHAPTER SIX

Environmental Justice and Detroit's Long Shadow

Michelle Martinez

Detroit Is Our Home, Our Toxic Home

At West Grand Boulevard and West Vernor sits the crossroads of the Latinx diaspora of Detroit, or at least how I have known it. The midway point between two Catholic churches, Ste. Anne's Parish and Holy Redeemer. As a girl, the Bowen library and the Honey Bee Market were the essential markers of our community. Detroit became the home for Mexican Americans during the Great Migration north, as laborers looked for work. By 1920 about 10,000 Mexican immigrants went north and landed in Detroit. One hundred years ago, the migration north was fueled by the desire for work, higher wages, freedom from slavery and civil war. My family was among the early migrants, working on the rail, selling canned goods, and shining shoes. They settled on the banks of the Detroit River, the northern border to Canada. We have been defined in many ways by this geography—our proximity to the borderlands, the Great Lakes, and the automotive industry. For five generations, my family has intermittently lived and worked within a two-mile radius, and like the diaspora itself, holding and recapitulating the landed memory of our people, fraught with the stories of movement across borders of place and identity.

Detroit is a changing landscape. As in many places across the United States, "border security" has dislocated the diaspora, pushing its boundaries away from its original core of SW Detroit to inner-ring suburbs. New immigrants coming from south of the border are locating in places that are more industrialized, in places where historically African American and working-class white people lived: in the shadows of factories.

Detroit is the quintessential story of environmental justice and injustice. For over a hundred years Detroit has been involved in the heavy industrialization and proliferation of the automotive industry and steel plants, and before that machinists, forges, steam engines, and more. But most importantly, conceptualizing Detroit, the Motor City, as the birthplace of the assembly line—singularly responsible for catalyzing the global dependence on the combustion engine—is critical within the context of climate change because of the existential threat now presented by its very existence. Detroit built the first highway; we exemplify the idea of suburbanization. At the height of the economy, post-NAFTA, an estimated 16,000 cars and trucks traveled over the international Ambassador Bridge every single day.[1] Buying food from a high-quality full-service chain grocery store requires driving to the only Meijer at the city limit.

Aside from producing cars and trucks, Detroit has the state of Michigan's only oil refinery, Marathon Oil. It receives tar sands—heavy oil sands from Alberta, Canada—through pipelines that traverse the Great Lakes. For over five decades the production of the automobile has been buttressed by the burning of Rocky Mountain coal and coal removed through the strip mining from the Powder River Basin, transported to the Great Lakes on rail. Since the late 1950s Detroit's energy system has been 80 percent coal and 20 percent nuclear. The reliance on coal has been dominated by one utility that was guaranteed a monopoly and guaranteed a rate of return on equity that hovers at about 10 percent. High return on investments has incentivized its transition to methane gas, fracked from the ground of the Marcellus Shale and piped from Pennsylvania, Ohio, and the Rocky Mountains. Fossil fuels have been the bedrock of the Detroit economy. The longitudinal distribution of the goods, and the externalization of their toxins, are ill-accounted for. As such the corporeal and geographic proximity to those toxins must be examined so as to account for the inequitable distribution of each on different communities. Only by reviewing the entire life cycle of the fossil fuel industry can we balance the accounts in the process of forming the new regenerative economy towards an era of radical survivance in the age of climate change.

For example, after the passage of NAFTA, truck traffic on the Ambassador Bridge rose exponentially. I remember as a teenager in the 1990s semi-trailer trucks transporting goods, idling on the highways for hours. Lines of rear brake lights twinkled day and night like lights on a Christmas tree through our community. We knew the increase of truck traffic was causing health issues. A public health scholar at the University of Michigan, Dr. Amy Schulz, tells of the socially disparate impact of pollution and increasing hospitalizations due to stroke and heart attack from roadway traffic. Those trucks that sat idling on the international bridge for hours released tiny diesel particulate matter exacerbating people's health issues, causing a spike in emergency room visits. The fossil fuel toxins released from combustion enter our airways and run through our veins. Particulate matter is so small it can break through the blood barrier—pushing people over the thresholds of asthma, stroke, and heart attack. Yet since the passage of NAFTA, household wealth has gone down, and health impacts from the transit associated with the movement of goods—and the restriction of free movement of people (mostly laborers) across the border—have gone up. We now have longitudinal data, our bodies, to account for the lived experience of long unaccounted for externalities.

The Black, Latinx, and Arab American populations that surround the industrial core have borne the brunt of health impact from industrialization as evidenced from elevated rates of asthma, heart disease, stroke, and lung cancer. Asthma is the number one reason for missing school in Detroit. In a time when we fight for equal education for per-pupil allotment, we are sending 84 percent of Black and 65 percent of Latinx children to schools in close proximity to contaminated sites, which impacts not only their health, but also their performance in school. This is not a situation that communities have "chosen," but the historic set of racialized geographies and economic factors have left communities of color geographically gridlocked in ways outside of their control. Our communities believe, in part, that the toxins and the associated health impacts are normal and/or necessary parts of our communities—when they are, in fact, the very reason for the degradation of our lands, homes, and bodies.

Now, I want to acknowledge that Black and Latinx people have only very recently migrated to Detroit or Waawiyatanong (the Anishinaabe term for Detroit, literally "where the water goes around"). And we cannot dismiss the very long-lived histories of Anishinaabe peoples whose residence on the land called Detroit over the last 14,000 years helped to maintain a healthy balance between the humans and other-than-human beings in the Great Lakes basin. The arc of environmental injustice has been much longer in formation for Anishinaabe peoples from the settlement of the lands in Western Ontario and Great Lakes Basin. Anishinaabe people are also contemporaneously impacted by toxins, notably in the case of Aamjiwnaang, where 40 percent of all residents have asthma as a result of Chemical Valley in Sarnia, Ontario. The significance of the impact on Anishinaabe peoples in Canada—like residents of Detroit—of living in the shadow is deeply embodied over many generations. Occupational therapist Jim Brophy talks about the impact on tours he conducts with activists like Lindsay Beze Gray, testifying that the persistence of endocrine disruptors in the land and water have resulted in the disproportionate birth of girls over boys, a chemically induced gender-ratio imbalance. Anishinaabe people, unlike the Black and Latinx diaspora, are living on their original lands, an inseparable relationship of history, kin, belief systems reinforcing that connection.

As such, Anishnaabek have been and continue to be impacted not only by the toxins, but also by the dislocation of what Dr. Kyle Whyte, Anishinaabe philosopher at Michigan State University, calls a continuous disruption of the "moral landscapes" within the settler framework beyond contemporary definitions of environmental justice:

> injustice also occurs when the social institutions of one society systematically erase certain socio-ecological contexts, or horizons, that are vital for members of another society to experience themselves in the world as having responsibilities to other humans, non-humans and the environment. Injustice, here, involves one society robbing another society of its capacities to experience the world as a place of collective life that its members feel responsible for maintaining into the future.[2]

Environmental justice for indigenous people challenges non-native peoples on Turtle Island to think beyond just the distribution of the goods, products, or material harm caused, but also the very existence of its constructs as formations that make other societal manifestations possible.

So what does environmental justice mean in the Motor City? I propose that environmental justice goes beyond equal protection under the law. The laws themselves are inadequate to protect health, ecosystems, and cultural and spiritual freedoms. They most obviously are not equipped to deal with the futurities exemplified by climate change. Instead, environmental justice must excoriate and repatriate the longitudinal aspects of justice—distributive, relational, spiritual, and corporeal—to restructure our economies in a way that rebuilds trust between peoples, restores the homes, lands, and sovereignty to nations, and regenerates the majestic splendor that has been devastated to sustain life on this earth. In this essay I will speak about my personal experience as a Latinx, Mestiza, Apache native in Detroit and the search for environmental justice in the age of climate change.

Forms of Justice

Typically, in the world of environmental justice, activists and advocates are fighting for two things: (1) a seat at the table and (2) a say in the decision that is made. Many people have heard the adage, "If you're not at the table then you are on the menu." This statement comes from the concept of procedural justice, the idea that if you are given a seat at the table with powerful decision-makers then you will be heard, you will be negotiating, you will be visible. Half the fight is just being seen, heard, or even believed. But there is a difference between having a voice and having a say. That difference is critical. The latter means possessing the *ability* to leverage and apply the political pressure necessary to deliver beneficial results. This is more than a notion because *by definition* environmental justice communities are vulnerable to toxins because of political disenfranchisement, systemic poverty, and racism. This latter issue is coined distributive justice: the idea that the goods (and harms) produced through industrial, economic, or colonial processes can be redistributed.

Much of the work in Detroit over the last ten to fifteen years has been committed to procedural and distributive justice. It is clear that the harms of the production of the automobile have resulted in disproportionate harm to Black, Latinx, Arab American, and Native peoples in southeast Michigan. We have four times the asthma rate of the rest of the state. According to activist Donele Wilkins, every single lot in Detroit was declared a brownfield—because they got so tired of counting *every lot,* the city itself was declared a brownfield. Southwest Detroit has a statistically significant higher level of bronchial and lung cancer. Detroit's children even have higher lead exposures than those in Flint, Michigan. The harms are corporeal, ecological, even spiritual. Toxins are everywhere—and in each study it is revealed that these harms are disproportionately impacting people of color. Which forces us, as environmental justice activists, to think again about justice and retrospectively consider the conditions that made this possible.

Despite the magnitude of the problem, we are just beginning to calculate the newest manifestations of harm from the fossil fuel industry, broadening now from just the life cycle of toxins to climate impacts. We must, therefore, broaden the lens of justice beyond procedural and distributive justice to *transitional* justice. This third form of justice rests upon the idea that *we can,* in fact, restore our bodies, restore these lands, restore relationships with peoples and nations harmed in the formation of the fossil fuel economy, and potentially heal from these bloodied histories through political reconciliation. Esme Murdock, Black feminist scholar and environmental justice philosopher, pushes us to review the limits of transitional justice before acquiescing and necessarily examine indigenous lands, relations, and the history of settlement beyond the liberal limits of reconciliation: "we must think again exactly what is being reconciled and what precisely we are transitioning to when the outcomes of reconciliatory processes are not transforming colonial socio-ecological systems and structures."[3]

So what does that look like? And can we do that before it's too late? Or is it too late already? We are being asked to take on the most critical existential crisis ever known by humanity. If you acknowledge and accept that climate change is already underway, and for many peoples the crisis is underway, you understand the urgency of this question. But the moral question is *how do we rebuild trust, relationships, mutuality to increase the potential survivance of many peoples, especially the most vulnerable before hedging against the cliff?* Survivance, a concept elucidated by Anishnaabe scholar Gerald Vizenor, is "more than survival, more than endurance or mere response. . . . Survivance is an active repudiation of dominance, tragedy and victimry."[4] Or are we in a paradox in which we are willing to subjugate the peoples already victimized by this system, in advancement of CO_2 reductions regardless of its detrimental outcomes for those most vulnerable?

The different forms of justice described here—procedural, distributive, and transitional—are not in competition with one another, nor are they comparative in the sense that if one practices one, they are not truly fulfilling the principles of environmental justice. But these forms absolutely cannot exist without one another; they must be in conversation to be accountable to a holistic future that centers and advances the connection between humans and other-than-human beings.

Detroit Case Study of Environmental Justice

The ZIP Code of 48217 is renowned as the most polluted community in Michigan; more pollutants are emitted there than anywhere in the state of Michigan.[5] It is at the most southern end of Detroit and primarily African American. 48217 resident Dr. Delores Leonard describes the settlement of Black soldiers to the area after World War II, as it was one of the few places where African American veterans were permitted to live and buy homes. In 2006, 48217 was negotiating a proposed expansion of a seventeen-acre Marathon Oil Petroleum facility, one that would process tar sands from Alberta.

Athabasca tar sands deposits are large, extremely heavy crude oil deposits located in northeastern Alberta, Canada. The Athabasca deposit in Alberta is the largest known reservoir of crude bitumen in the world and the largest of three major oil sands deposits in Alberta, making Canada the largest producer of oil outside Saudi Arabia and Venezuela.[6] The tar sands are piped through the Canadian Prairies, cross into the United States in northeastern North Dakota and travel across northern Minnesota, Wisconsin, and Michigan's Upper Peninsula before heading south toward Detroit. That has great significance. Enbridge's Line 5 travels under 89 percent of the United States's surface fresh water, the Great Lakes. Built in 1953, the pipeline has released over 1.1 million barrels into the watershed over its lifetime. Our precious watershed is nearly a quarter of the Earth's surface freshwater, a critical resource for fragile ecosystems, and a spiritual home for over 14,000 years, and it will be a precious resource in the coming era's unfathomable scarcity. Michigan is also home to Enbridge's Line 6B, which erupted in 2010, causing the nation's largest inland oil spill, resulting in 20,000 barrels of crude oil pouring into the Kalamazoo River. Both Enbridge Line 5 and Line 6B are linked to the Marathon Oil refinery.

Residents of Southwest Detroit were concerned about the impacts of additional pollutants in the neighborhood. The community is already overburdened by industry. The area of Southwest Detroit and

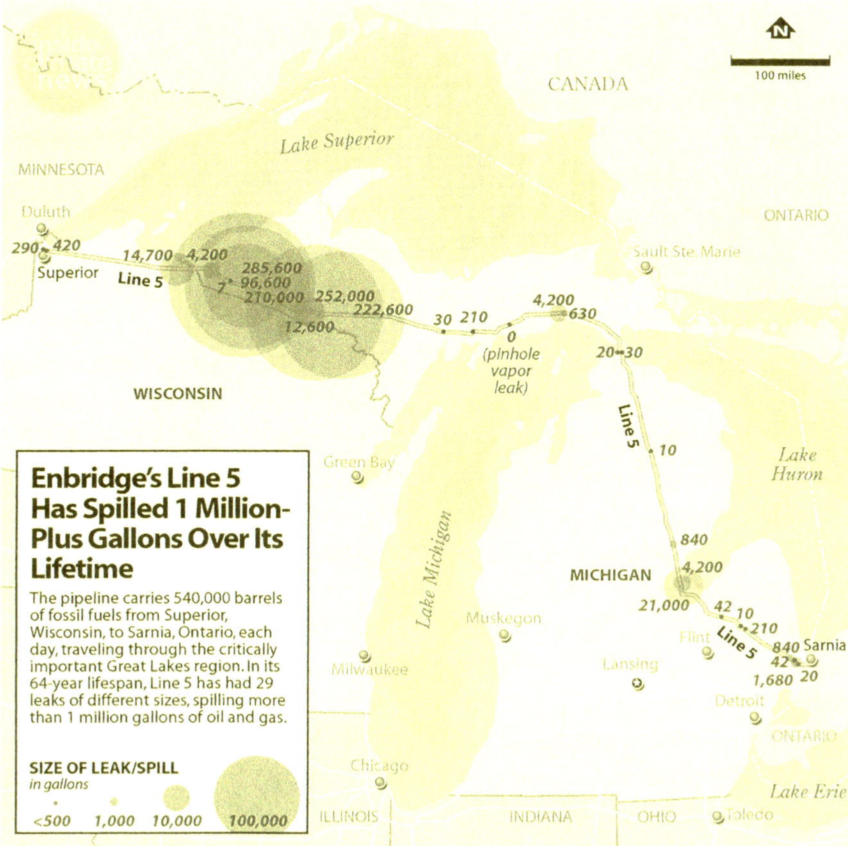

FIGURE 2. Gallons spilled from Enbridge Line 5 (*Source*: National Wildlife Federation. Image by Paul Horn, InsideClimate News)

neighboring South Dearborn produces cars, steel, and chemicals; refines petroleum; processes human waste; and burns coal and methane gas. The auxiliary storage of the waste products and its transports result in some of the heaviest and persistent pollutants. The goods and services of those industries are mined, piped, and transported across the border to Canada and the Western United States on rail, ship, and heavy diesel transit.

And the residential community of 48217 is directly adjacent to the industrial sector. There is absolutely no barrier between schools, playgrounds, churches, recreation centers, and industry. The outflow is directly processed and discharged into the River Rouge, a tributary of the Detroit River, or directly into the Detroit River.

In 2007, I attended a meeting in the community of 48217 as a recently graduated student and intern with the Sierra Club. Environmental Justice organizer Rhonda Anderson had told me to go down to 48217 and help where I could. This particular meeting was critical because it echoed what environmental justice communities all over the country were asking for—a fix to the problem of the longitudinal effects of industrial contaminants over the lifetimes of two or three generations. And the problem was not just of one facility, but of several dozens.

Detroit is a case where industrial production has seen a rise and fall, where the automotive industry peaked in the mid-twentieth century. But in the period just before the economic crash in 2008, and resurging in the 2010s, a doubling down on the industrial core compounded toxicity issues. In the late twentieth century, many of these facilities had closed and left superfund legacy sites, some of which were grandfathered into old laws that were not protective of health standards. Other facilities were either renovating and expanding—like the case of Marathon Oil. Previous to the economic collapse of 2008, many heavy industries in Detroit were conducting massive expansions, still riding on the coattails of NAFTA and lax regulatory regimes. The Detroit International Crossing, a second bridge to Canada over the Detroit River, was proposed and approved in the Southwest Detroit neighborhood of Delray. Hundreds of people showed up to testify at the Environmental Impact Statement public hearing for the crossing, yet the concerns of community about their respiratory health fell on deaf ears. Instead regulatory agency heads argued pollution would go down based on their models. Their models erroneously looked at tri-county regional transit flows—and not local congestion.

The Detroit Intermodal Freight Terminal, which proposed an expansion of freight that would combine three regional exchange terminals into one in the residential communities of Southwest Detroit, was approved but remains unfunded. And just forty miles downriver, a $2.2 billion expansion of the DTE Energy coal-fired power plant was approved for the energy monopoly utility headquartered in Detroit, making it one of the largest coal facilities in the United States—and it was permitted to burn tar sands waste, known as petroleum coke, for energy. Notably, DTE Energy received approval in 2017 for a new methane gas burning power plant forty miles up the Detroit River in St. Clair County. Yet the regulatory regime outlined in the Clean Air Act treats each facility as if it were in a vacuum. It grants permits based on a pollution budget. The ambient air quality monitoring system, regulated by the state, is established according to federal law. The problem? It is woefully inadequate technologically, systematically underfunded, and underreported, as companies are also responsible for self-reporting their own environmental monitoring.

Unfortunately, the Marathon Oil facility came online in 2013—despite broad protest—yet that meeting in Southwest Detroit laid the political foundation for a long-term movement toward environmental justice. The community of 48217 is strong and organized, and it has gained a hard-earned seat at the table for environmental justice. The community still negotiates with the company directly, receiving several benefits like a new recreation center, filtration systems, and regular meetings with the company. For over a decade, people like Theresa Landrum and Dr. Delores Leonard have had to fight tooth and nail for their dignity. This advocacy has not stopped the pollution—and, in fact, the safety at the facility is rather poor, resulting in multiple illegal releases and a facility fire since opening. But this movement has resulted in another solution, what researchers now call the *cumulative impact* of pollution. I believe that meeting in 2007, where residents ended up screaming at the state agency, finally made possible a space for cumulative impact [CI] analysis. CI goes beyond the limits of the Clean Air Act, and it combines the ideas that pollution exposure, health outcomes, and sociological variables like race, poverty, education, and language accessibility can geographically map where the impacts of environmental injustice are the highest, and compare those with other areas.

So What? You Chose to Live There!

Richard Rothstein's *The Color of Law* (2018) very clearly links the historic markers of housing segregation to the history of slavery. "In 1866, Congress enforced the abolition of slavery by passing a Civil Rights Act, prohibiting actions that it deemed perpetuated the characteristics of slavery." At this point, housing was included in the ban, but in 1883 the Supreme Court rejected the notion that housing markets could be a badge or incident of slavery, and so housing segregation was supported and enacted by governmental policies, social norms, and banking institutions. Until 1968, when the Supreme Court overturned that decision in *Jones v. Alfred H. Mayer Co.*, in which a developer refused to sell a home to Mr. Joseph Lee Jones in St. Louis solely based on race, the parameters of the Fair Housing Act were unregulated.[7] Detroit was not exempt from the implications of racism in housing policy and the downstream effects on their localities. The homes and the areas available to postwar Black families were directly adjacent to the industries we have discussed. And the combination of anti-Black racism, housing discrimination through redlining, and workforce inequity resulted in many of the dynamics that have compounded into environmental injustice and racialized geographies, making Detroit one of the most racially segregated metropolitan areas in the United States today.

Data available from the 2015 American Community Survey reveals the extent of this segregation in the Detroit Metropolitan Area. Though housing segregation prevented many families from moving into affluent suburbs forty to fifty years ago, it also codified *generations* of people into a racialized geography. At the same time that Detroit experienced "white flight," it also was hammered by urban disinvestment and capital flight, culminating in the collapse of the municipal financial system, marked by Detroit's 2013 municipal bankruptcy—the largest municipal bankruptcy in American history. Those who were fortunate to maintain any personal wealth while living in Detroit in the latter end of the century experienced a multitudinous loss in property values, intergenerational wealth, opportunity, quality of education, and public safety. According to the US Census Bureau, the median value of owner-occupied housing in Detroit from 2014 to 2018 was $48,700. That is the median. Many homes (upwards of 24,000, in fact) went into foreclosure either by the bank or the county and were valued at only $500 in 2015. Compare that with Ferndale, Michigan, a historically working-class white neighborhood just north of the city's limit of 8 Mile Road, whose median home value was $148,100 for the same period.[8]

Cumulative Impact [CI]

As communities in Detroit were asking for a respite from the pollution that accounted for past harms, so were other communities all over the United States. The infrastructure built or expanded in the postwar era across the United States is now in decline. Many of the pipelines, power plants, and production facilities have reached the end of their life cycle. The roads, bridges, highways, and waterways are at a crossroads. And the question that continues to plague us is: What system will we build and how will we pay for it? The question is under extreme scrutiny by thought leaders and those victimized by environmental injustices and its ravages. In a field hearing, Senator Gary Peters listed the costs associated with the 2014 flooding around Detroit as more than $1 billion. In Warren, an estimated 18,000 structures—about one in three homes—were damaged. In 2017, Isabella County flooding caused about $90 million dollars in damage. More than $20 million worth of crops were lost. In Michigan's Upper Peninsula last year, flooding in the Houghton area caused nearly $100 million in damage to public infrastructure.

Some areas of Michigan have seen more rain in 2019 than ever before recorded. Environmental justice advocates are now taking up space in the climate arena, insisting that *past* harms cannot be ignored as we plan for *future* systems. Cumulative impact [CI] is one tool that environmental justice activists and advocates have been developing to demonstrate how people of color and indigenous communities should be first in line for the investments of the future, as we have borne disproportionate impacts of the fossil fuel economy.

In 2017, the Michigan Environmental Justice Coalition began working with scholars at the University of Michigan's School of Environment and Sustainability to complete a study on the cumulative impact of environmental harms on communities of color across the state. We knew that, after five years of the Flint Water Crisis, the expansion of the fossil fuel industry on the Great Lakes and the magnitude of the climate crisis needed to be addressed systemically. The first step in any climate policy *must be* cumulative in its impact analysis. The ability to distribute the goods of any system requires the identification of the harms. In California, cumulative impact is a policy that is codified through CAL Screen. It combines pollution exposure and socioeconomic data to create maps of the hardest hit communities. California, for instance, uses this methodology to direct investments for clean energy technologies. The researchers at the University of Michigan not only interviewed environmental justice advocates but conducted the first statewide assessment of environmental injustice in Michigan.

Results indicated that when you combine pollution burden with population characteristics you create an environmental justice burden index that is comparable across the state. Urban areas, such as Detroit, Grand Rapids, Flint, Saginaw, and Lansing, show how environmental justice hotspots are present throughout the state. Both qualitative and quantitative data demonstrated that environmental goods and harms are inequitably distributed in the state of Michigan:

> Vulnerability is exacerbated by instances of procedural injustice, such as decision-making by emergency managers instead of elected leaders; distributive injustice, such as income earned from energy production leaving communities; and corrective injustice, such as lack of compensation for damages. A long list of barriers, such as lack of transportation and feeling ignored by the government, hold back community members' efforts of improving the health of themselves and their communities.[9]

Interviews revealed that environmental justice leaders characterized affected communities as resilient despite adversity. The network of social justice advocates and activists is the social capital that makes the leadership and political potential possible. Others cited a myriad of environmental, social, and health impacts, including lack of air and water quality, gentrification, asthma, and cancer. Those interviewed also spoke to the psychological impacts not only of being personally affected or witnessing these issues, but also of fighting what seemed like "an insurmountable battle for justice."[10] Additionally, participants described numerous barriers to achieving environmental justice, most notably lack of political will and the erosion of democratic processes. Participants disclosed utilizing a variety of tools to advance environmental justice and expressed a desire for a Michigan-specific tool to measure cumulative impacts. However, they expressed that the tool must be accompanied by strong state-level environmental justice policy in order to be effective.

The quantitative aspects included quantifying eleven environmental indicators: air toxics cancer risk, air toxics respiratory hazard index, diesel particulate matter (PM), ozone level, PM 2.5 level, traffic proximity and volume, lead paint indicator, proximity to National Priority List sites, proximity to risk management plan facilities, proximity to treatment storage and disposal facilities, and a wastewater discharge indicator. The six social indicators used in the study were: percent minority residents, percent of households living below twice the federal poverty level, unemployment rate, percent of residents with less than a high school education, percent of households living in linguistic isolation, and percent housing-burdened low-income households.[11] By combining social and environmental indicators and placing them on a map, we can see very clearly that racialized geographies are present in relation to exposure to and impact from environmental toxins—yet the political capital to course-correct seems insurmountably unattainable.

Overall the study also found that when you layer the environmental and social data that is publicly available, the top 5 percent most polluted areas are composed of 81.5 percent people of color, compared with 25.1 percent of the general population.[12] This is a classic environmental justice issue, proving that environmental racism is alive and well in Detroit. But the limitations of this study are that we cannot and have not been able to access other critical data that more broadly addresses the impact of environmental justice. For example, exposure to chemical fertilizers within farm worker populations is not characterized here. The violation of sacred sites in mining on indigenous landscapes are not included. This study is not the panacea we all hoped for, but it provides an opportunity to understand and codify the distribution of harms, so that we may repair the trust and relationships—and perhaps increase survivances—in the future we all steward.

Identification of environmental injustices using a tool like cumulative impact assessment is a critical first step in the creation of any environmental or climate policy. If we cannot identify who has borne the impact and where they live within the current system, we cannot understand the vulnerabilities set in place as the climate crises unfold, nor can we establish the boundaries of the creation of a new economy that addresses the historic harm to systematically dismantle the patterns of injustice for the future. We do not pretend this will solve all of the problems associated with the climate crisis. And, as we know with some certainty, each and every geography has a different and unique set of histories and economies at play when discussing adaptation or mitigation. But this tool does provide us a framework from which we can engage the question of racially disparate impact.

The wicked question before us, then, is after identifying those who have been impacted, how can we create policies that would first limit the continued pollutants and then invest in renewable energy structures that invest wealth back into communities, divesting from climate-changing industries. Different places all over the country are turning towards retooling our economy, building workforce programs for veterans and returning citizens, investing in solar in schools to offset education costs in public schools, decentralizing energy systems and creating resilience centers, and tying affordable housing financing to clean energy credits—all wealth building opportunities that will alleviate and eliminate poverty *and* the climate crisis. Our movement is solutions-rich and resource-poor. Figuring out how to actualize the economy through local food systems, clean affordable transit, and off-grid solutions is proliferating. But our finance systems are falling behind; in fact, they are actually intensifying the climate crisis. More troubling, they are denying the conditions which facilitate the trust needed to heal historic harm and reconcile the deeply unequal fastenings of the settler state.

A recent report created by the Rainforest Action Network, Bank Track, Sierra Club, OilChange, Indigenous Environmental Network, and Honor the Earth indicates that thirty-three banks all over the world have invested upwards of $1.9 trillion in the fossil fuel industry since the Paris Agreement—each year investing incrementally more. That is more than all of the currency in circulation in the United States. The worst of those banks are the US-based JPMorgan Chase, leading all others by almost 30 percent. JPMorgan Chase, Wells Fargo, Citibank, and Bank of America represent the top four. These banking investments are not in a vacuum. The United States *direct* governmental subsidies are reported to be exceeding the budget of the Pentagon. At a time when we should be investing in solutions through an equity lens, banking in the United States is fueling the climate crisis in very literal ways.

Conclusion

The compounding evidence of environmental racism has been proven globally—and locally—over and over again. It is true in Michigan. It is true in Detroit. The data contends and supports the claims made by communities over the last decades and beyond—that the cumulative and longitudinal impacts of industrialization are wearing at the quality of life and life itself. There are efforts toward identification of those hardest hit by these toxic inequities, which have manifested over the life cycle of the fossil fuel industry. There are also policy and technological solutions to alleviate current conditions and ameliorate future harms.

But the growing and indisputable truths of an unfolding climate disruption are exacerbating political and socioeconomic inequities at an even more alarming rate. When you begin to uncover the historic underpinnings of the systems motivated by profit and disinvestment, it is heartbreaking. When we sit at home and read headline after headline about the devastation from typhoons in Japan, flooding in the Philippines, hurricanes in the Caribbean, droughts and hunger in the Middle East, historic rainfall or crop loss, an undercurrent of catastrophism takes over. What is it that we can do? These questions are not new. Disruption, dislocation, and unsettled deterritorialization are and have been the underaccounted for or totally erased history of people of color.

We know there are pathways out of the crisis; we *must* hold that vision. There are ways of being that will alleviate the crisis we are in, so we must take bold political action to make them possible. There is an abundance of compassion, yet it takes a wildly controversial form of love for others beyond borders and cages. There is a wave of bravery and perseverance giving everyone who believes that human beings—regardless of their race, income, or national status, regardless of their gender identity, their physical or cognitive abilities, or the God to whom they pray—*all* have a right to live, *all* have a right to migrate and drink water, and *all* have a right to participate in restoring our relationships and our communities. We must *believe in the right* to rehabilitate our connection to nature, to repatriate lands to peoples and reconcile those ways of knowing across difference—but it takes everyone. Because even if *every* single human in the United States spent every single penny they have on solutions, it would not amount to the total that banks have given the greedy and catastrophic fossil fuel industry in the last three years alone. It will take more than money. We must collectively learn how to be totally fearless in defending the political potential of life and its restorative powers, because the solutions themselves can *only* be born from the immaterial and magical moment of unknowing created when two minds, different ways of being, and worlds come together to restore the damage of segregation and colonization. It is happening all around us. All you need to do is look, and then get to work.

Notes

1. Andrew Clement, Krista Boa, Simon Davis, and Gus Hosein, "Towards a National ID Card for Canada? External Drivers and Internal Complexities," in *Playing the Identity Card: Surveillance, Security, and Identification in Global Perspective,* ed. Colin J. Bennett and David Lyon (New York: Routledge, 2008), 248.

2. Kyle Powys Whyte, "Indigenous Experience, Environmental Justice and Settler Colonialism," in *Nature and Experience: Phenomenology and the Environment,* ed. Bryan E. Bannon (Lanham, MD: Rowman & Littlefield, 2016), 158.

3. Esme G. Murdock, "Storied with Land: 'Transitional Justice' on Indigenous Lands," *Journal of Global Ethics* 12, no. 2 (2018): 236.

4. Gerald Vizenor, *Fugitive Poses: Native American Scenes of Absence and Presence* (Lincoln: Nebraska University Press, 2000), 15.

5. Kyle Powys Whyte, "Indigenous Food Systems, Environmental Justice, and Settler-Industrial States," in *Global Food, Global Justice: Essays on Eating under Globalization,* ed. Mary C. Rawlinson and Caleb Ward (Newcastle: Cambridge Scholars, 2015), 159.

6. Kevin P. Timoney, *The Peace-Athabasca Delta: Portrait of a Dynamic Ecosystem* (Edmonton: University of Alberta Press, 2013), 355.

7. Richard Rothstein, *The Color of Law: A Forgotten History of How Our Government Segregated America* (New York: Liveright, 2018), viii, ix.

8. "QuickFacts: Detroit city, Michigan," US Census Bureau, https://www.census.gov/quickfacts/detroitcitymichigan.

9. Laura Grier, Delia Mayor, and Brett Zauner, *Assessing the State of Environment Justice in Michigan* (Ann Arbor: University of Michigan School for Environmental and Sustainability in Partnership with the Michigan Environmental Justice Coalition, 2019), 111.

10. Ibid., 3.

11. Ibid., 4.

12. Ibid., 6.

CHAPTER SEVEN

Beyond the Prophetic Temptation of Ecological Disgust

Rudolph P. Reyes II

The question that animates this chapter is one asked by writer, poet, and activist Eli Clare: "How do we witness, name, and resist the injustice that reshape and damage all kinds of bodyminds—plants and animal, organic, and inorganic, nonhuman and human—while not equating disability with injustice?"[1] There are a variety of responses to this question. One is a nondisabled response. This response may be one of puzzlement because of the belief that disability is something to be avoided and hopefully one day cured. Another one is a disabled response. This response understands the question differently: How can I witness, name, and resist injustice without hating myself? Or put another way, how do I resist injustice without supporting ableist ideologies? Disability liberation is not liberation from particular embodiments, but rather liberation from a disabling society.[2]

Disability is a nexus for several forms of structural inequality and the intersection of race and disability compound it. There are higher rates of poverty for disabled people of color than nondisabled people of color. The poverty rate is 11 percent higher for disabled Latinxs than for nondisabled Latinxs.[3] Similarly, the poverty rate for disabled African Americans is 17 percent higher than for nondisabled African Americans.[4] Disabled youth are overrepresented among incarcerated youth with 85 percent having a disability and youth of color overrepresented in this population.[5] Only 37 percent of these youth received accommodations in school.[6] The probability of adult disabled people of color getting arrested is higher than nondisabled people of color.[7] The probability for disabled Latinxs is 15 percent higher—and for disabled African Americans is 18 percent higher—than their nondisabled peers.[8] Disabled people of color are greatly impacted by structural inequality, but often left out of movements to resist poverty, police brutality, and incarceration.

For disabled people of color, racism and ableism intersect in a myriad of ways. It is challenging for people of color to identify as disabled or neurodivergent—a positive political identity of neurological difference—because it adds a stigmatized identity upon an already marginalized one. Disabled people of color embody a "qualitatively different" position. What dis/ability critical race scholars David Connor, Beth Ferri, and Subini Annamma note for disabled students of color is also true for disabled people of color generally: "The label of dis/ability situates them [disabled students of color] in unique positions where they are considered "less than" white peers with or without dis/ability labels, as well as their non-disabled peers of color."[9] This qualitative different positionality requires a distinct form of liberative analysis.

I approach ecojustice from a liberative ethical framework and a critical disability studies approach. As a neurodivergent Chicano, I know there is no ecojustice without both racial and disability justice. My neurodivergence is my birthright; I did not acquire my disability. This difference requires me to listen to those voices who have acquired disabilities in wrestling with the question of resisting injustice without equating disability with it. The main argument of this chapter is for the reframing of the problem of water pollution because of the use of the prophetic temptation of ecologic disgust to mobilize action. To narrow my focus, I concentrate on lead as a form of water pollution.

My goal in this chapter is for those fighting for ecojustice to struggle with Eli Clare's question of witnessing, naming, and resisting injustice without making disability itself the injustice. I link Clare's critique of environmental awareness campaigns with ecocritic Sarah Jaquette Ray's concept of ecological disgust. I find that what Clare critiques is a form of ecological disgust. This ecological disgust is one manifestation of prophetic temptation. I begin with an examination of this prophetic temptation within the context of spreading awareness of the dangers of lead in drinking water. Then I point to ways to move beyond the prophetic temptation of ecological disgust.

The Prophetic Temptation of Ecological Disgust

Prophetic temptation is the temptation to use a shared dominant position in the struggle for liberation. I call this a *prophetic* temptation because it is a perennial temptation of progressive and liberative efforts to sacrifice other communities because of the expediency and explanatory power of leveraging that shared dominance.[10] An example is the suffragette movement's abandonment of women of color to leverage a shared whiteness to further a (white) women's right to vote. For environmental movements, this has meant the temptation of sacrificing people of color and people with disabilities through leveraging both a shared whiteness and abledness. Mainstream environmentalists, when engaging in this prophetic temptation, have obscured environmental racism and contributed to eco-ableism. This prophetic temptation appears in efforts to combat lead in drinking water.

One tool in these efforts is what ecocritic Sarah Jaquette Ray calls ecological disgust. Ecological disgust is a bodily discourse that makes a correlation between the body and the environment. It tells us that through our very bodies we can make a better environment by the food we eat, the clothes we wear, and the places we play. As Ray points out, ecological disgust has an othering effect.[11] The effect of this discourse is to divide people into two categories, ecological subjects or what she terms ecological others. Ecological subjects are those who are empowered to make certain consumer choices or action to protect the environment. Ecological others are those perceived as destroying the environment. The separation between ecological subjects and ecological others is not based on impact on the environment per se, but rather the embodiment of their respective group. For example, the fat body becomes an ecological other because it comes to represent a discount between the body and the environment, as well as a symbol of aversion. For Ray this bodily discourse is built off of the disabled body, where the disabled body represents humanity's divorce from nature.[12] This is evident in the metaphorical imagination which employs illness and disability metaphors to describe environmental destruction. In addition, ecological disgust aides in the masking of oppressive projects through the cloak of ecological concern. This is most evident in the way this bodily discourse of purity and dirt are used to reinforce racial ideologies of equating whiteness with cleanliness and purity and nonwhiteness as dirty and polluted.

I turn to an ongoing campaign to remove lead from drinking water in schools to uncover the ways this prophetic temptation plays out. I focus on digital media used to mobilize action towards this goal.

Environment America and the US Public Research Group launched the "Get the Lead Out" campaign. This campaign is targeting state and local officials to address lead in water in schools. At the heart of "Get the Lead Out" is mobilizing people to address this crisis by focusing on the harm done to children. Lead is a neurotoxin and does severe damage to bodyminds. There is a discrepancy in the ways these harms are described in this campaign.

On the landing page for the campaign, there is a banner with the image of a young child drinking from a water fountain. The image signifies the danger of children drinking polluted water in a place they should feel protected. The first paragraph below the header reads: "Medical experts estimate that more than 24 million American kids will lose IQ points due to lead exposure. Our research finds many of these kids will be exposed to lead in their schools' drinking water. That's why we're working to Get the Lead Out."[13] The message is clear, the potential harm to children is in the loss of intelligence.

On the campaign's back-to-school toolkit page is a slider with Frequently Asked Questions. Under the question "How does lead harm children?" a quote from the Environmental Protection Agency reads: "In children, low levels of [lead] exposure have been linked to damage to the central and peripheral nervous system, learning disabilities, shorter stature, impaired hearing, and impaired formation and function of blood cells."[14] Here we have a longer list of harmful effects. In this list the only effect that is given a diagnostic category is that of learning disabilities.

Environment America and US Public Research Group make available a "Get the Lead Out" report. This report provides more detailed information on the problem of lead, how it happens, and ways to get involved. The section "Lead in Schools' Water: A Threat to Children's Health" contains most of the information contained elsewhere, but with additional visuals. One visual is entitled "Figure 1: Adverse Effects of Lead Levels."[15] The figure is a drawing of a vile of blood with red blood cells and gray cells labeled PB for lead's chemical element. There is a scale with the amount of lead exposure corresponding with the effect on the bodymind. It starts at 10 μg/dl (micrograms per cubic deciliter) with development of toxicity and ending just below 150 μg/dl (micrograms per cubic deciliter) with death. The visual aims to demonstrate that the entry level for lead poisoning causes developmental delays. This visual presents the main thrust of the "Get Lead Out" campaigns to focus on the harm done to children by lead. The emphasis on the landing and toolkit pages plus the report emphasize that loss of intelligence is the primary concern driving environmental action to stop lead in schools.

There is, however, a discrepancy between the material that campaign uses to mobilize people towards environmental action and the resources given towards that mobilization. This is demonstrated in the sample material to share on social media and sample letters to the editor. For example, four sample tweets are given and only one refers to the harm done to children. The sample tweet reads: "There's no safe level of lead. As our children head #backtoschool, get the facts on safe drinking water from @EnvAm @uspirg."[16] Participants are also encouraged to share a video on the effects of lead on children. The video uses the following language: "It affects how our children learn, develop, and behave."[17] The language of learn, develop, and behave or learn, grow, and behave are used throughout the material to share from sample letters to the editor. While some might argue that the language of learn, grow, behave are shorthand for the effects of lead on children described elsewhere, the language of IQ is already a shorthand that is mostly absent from these advocacy tools.

"Get the Lead Out" campaign puts forward—as Eli Clare notes of similar efforts—a disability argument. The disability argument put forward constructs the problem of water pollution to that of children becoming disabled. The solution is to prevent disabilities by removing lead from schools. This disability argument is an example of eco-ableism that obscures environmental racism. Eco-ableism is the environmental discrimination against disabled people through the construction of environmental problems and means of fixing these problems. The ecological disgust of a disabled body is assumed and leveraged to combat water pollution. I am not arguing that we should not witness and name the slow violence done to children through water pollution, but rather reframe the narrative where disability is not the center of these acts of witnessing and naming injustice.

Moving Beyond This Prophetic Temptation

Prophetic temptations are contextual, as are the ways to move beyond them. There is a need for constant guarding against it or at least recognition when temptation is indulged. In combating environmental ableism and water pollution there are at least two ways to prevent falling into temptation. One is more general, and another is more specific. On a general level, this requires collaborating with disability communities and listening to those with acquired disabilities. On a more specific level, this requires embracing the neurodiversity paradigm. This does not encompass all the ways to move beyond the prophetic temptation, but rather is a starting point.

Writer and activist Elsa Sjunneson-Henry demonstrates this tension in her opinion piece entitled "Rubella gave me a disability. This is my message to anti-vaxxers." Sjunneson-Henry was born with Congenital Rubella Syndrome as a result of her mother contracting German measles while pregnant with her. Her mother knew neither that she needed to get an MMR vaccine booster nor that she was pregnant. Sjunneson-Henry notes that when she hears of a measles outbreak, she is caught in this tension between advocating for people to take MMR vaccine to prevent her disability and her advocacy understood as self-hatred of her own disability. Yet, she goes on to write:

> Threading the needle of not hating myself or my disability but finding the source of my disability utterly reprehensible has been a difficult one, but I've found it. I'm a strong believer in the need for the immune herd. I love the life that I have. But I don't believe that anyone should have to live a disabled life—let alone lose a baby—because of bad science and a wrongheaded disbelief in vaccination.[18]

Sjunneson-Henry advocates for vaccinations to prevent disease and protect those who are unable to receive vaccination. The fears of vaccinations are that it can cause autism. "What that means," Sjunneson-Henry explains, "is that people are willfully putting others at risk of death because they think autism is a worse fate—and therefore worth that risk that others might die."[19] Dialoguing with those with acquired disability upholds the disability justice principle, nothing about us without us.

As the disability argument in the previous section focused on the prevention of particular types of disability (namely, neurological difference), moving beyond the ecological disgust requires a shift to the neurodiversity paradigm. The neurodiversity paradigm as laid out by Nick Walker has three fundamental principles. First, *neurological difference is a valuable form of diversity.*[20] It is hard to dispute that humanity does not encompass a variety of neurocognitive styles. The neurodiversity paradigm goes beyond this fact to claim that this diversity is valuable. This is in contrast to other paradigms, where neurological difference is acknowledged but targeted for eradication. These varieties of neurocognitive styles are called neurotypes.

Second, *there is no normal neurological style.*[21] This second principle makes explicit that there is no normative neurotype by which others are judged. Neurotypical is the dominant neurotype and thus the neuromajority. Those who are neurodivergent are neurominorities. Third, *similar social dynamics of oppression and creative resistance that operate for other forms of diversity operate for neurodiversity.*[22] This principle does not equate one form of oppression with another. Every form of oppression has its unique history, genealogy, and experience. In arguing that oppression of neurominorities has similar social dynamics is to argue that neuro-ableism is structural. Neuro-ableism is not interpersonal malice. These three fundamental principles are necessary and interconnected. If one acknowledges that neurodiversity was valuable, but accepted that there was a normative neurotype, it would allow neuro-ableism unbated. Neuro-ableism can absorb the first two principles without accepting the third.

This is only the starting point in the process of moving beyond the prophetic temptation of ecological disgust. It allows for the urgent work of freeing our imagination in constructing urgency of action without making disability itself the injustice.

Conclusion

To witness, name, and resist the injustice done to Black and Brown bodyminds by water pollution requires an analysis of the intersection of environmental racism and water as my fellow contributors to this book have done. It also requires the witnessing and naming of the prophetic temptation of ecological disgust that constructs disability arguments to mobilize resistance to water pollution. To reframe these arguments for mobilization requires moving beyond this prophetic temptation by dialoguing with disabled people of color and embracing the neurodiversity paradigm. The failure to recognize the intersection of racism, disability, and ecojustice is to engage in a form of prophetic temptation that further marginalizes disabled people of color.

Notes

1. Eli Clare, *Brilliant Imperfection: Grappling with Cure* (Durham: Duke University Press, 2017), 56.

2. I do not subscribe to pure anti-cure disability politics because disability is a spectrum. As neurodivergent I cannot prescribe someone with chronic illness not to seek a cure or lessen their pain because I am anti-cure for my own neurodivergency. However, I do make a distinction between the need for some to seek cures for their impairment or chronic illness and pain from the political goal of disability liberation.

3. Nanette Goodman, Michael Morris, Kelvin Boston, and Donna Walton, *Financial Inequality: Disability, Race and Poverty in America* (Washington, DC: National Disability Institute, 2017), 12.

4. Ibid.

5. National Council on Disability, "Breaking the School-to-Prison Pipeline for Students with Disabilities," June 18, 2015, https://www.ncd.gov/publications/2015/06182015.

6. Ibid.

7. Erin J. McCauley, "The Cumulative Probability of Arrest by Age 28 Years in the United States by Disability Status, Race/Ethnicity, and Gender," *American Journal of Public Health* 107, no. 12 (2017):1978.

8. Ibid.

9. David J. Connor, Beth A. Ferri, and Subini A. Annamma, eds. "Dis/Ability Critical Race Studies (DisCrit): Theorizing at the Intersections of Race and Dis/Ability," in *DisCrit: Disability Studies and Critical Race Theory in Education,* Disability, Culture, and Equity Series (New York: Teachers College Press, 2016), 13.

10. My use of "prophetic temptation" differs from legal scholar David Skeel's usage of prophetic temptation to describe a tendency among Hauerwasians to refrain from extended participatory engagement in favor of short-lived prophetic critique. As used here, however, the prophetic temptation leverages a shared dominance in the struggle for liberation because of the expediency and explanatory power of using that shared dominance to achieve a liberatory goal.

11. Sarah Jaquette Ray, *The Ecological Other: Environmental Exclusion in American Culture* (Tucson: University of Arizona Press, 2013), 5.

12. Ibid., 9.

13. Environment America, "Get the Lead Out" (2019), https://environmentamerica.org/feature/ame/get-lead-out.

14. Ibid.

15. John Rumpler and Laura Deehan, *Get the Lead Out: Ensuring Safe Drinking Water for Our Children at School* (Denver: Environment America Research & Policy Center, 2019), 4.

16. Environment America, "Get the Lead Out: Back-to-School Toolkit" (2019), https://environmentamerica.org/feature/ame/get-lead-out-back-school-toolkit.

17. Ibid.

18. Elsa Sjunneson-Henry, "Rubella Gave Me a Disability. This Is My Message to Anti-Vaxxers," *CNN*, February 19, 2019.

19. Ibid.

20. Nick Walker, "Throw Away the Master's Tools: Liberating Ourselves from the Pathology Paradigm," *Neurocosmopolitanism: Nick Walker's Notes on Neurodiversity, Autism, & Self-Liberation* (blog), August 16, 2013, http://neurocosmopolitanism.com/throw-away-the-masters-tools-liberating-ourselves-from-the-pathology-paradigm.

21. Ibid.

22. Ibid.

WATER AS WEAPON

CHAPTER EIGHT

Praying for the Water—Racial and Religious Solidarity in the Shadow of Coal Ash

Eric C. Smith

Ash, Soot, and Lies

The backyard of the house I grew up in looked out over a forest of pine and hardwood, ringed with kudzu where the trees met the yard, and above that, interrupting the horizon, stood two gray smoke-stacks. The towers belonged to a coal-burning power plant, the Belews Creek Steam Station.[1] The people in my small rural North Carolina town had an uneasy relationship with this power station. It was an employer in an area where there was not much to do for work if you didn't work in tobacco. But also, every morning, we would walk out of our homes and find a fine layer of soot—fly ash, as it is called—covering everything. You could run your finger over the windshield of your car and come up with a smudge on your fingertip.

This was the 1980s, still the early days of the environmental movement, but there was enough concern and awareness that the power company that ran the plant had started making a big deal about how much it had invested in pollution controls for that plant and those towers. There was new technology in them, we were told, technology that made them safer, scrubbers that reduced the smoke and the ash that came out. But still we came out every morning to the soot.

Beginning when I was about ten years old, my father worked at that power plant. He was a nighttime security guard who sat in a guard booth and regulated traffic in and out of the plant, including the long trains that came to disgorge tons upon tons of coal to be burned. Years later, not long before he died, my father told me that when he had worked there, he had learned the power company had been telling the truth about that pollution technology. They had installed it, and they did run it—all day long. But at night, when no one could see the black smoke against the night sky, they turned off the pollution technology and just let the smoke and ash go up into the air. It was expensive to run the pollution controls. It reduced the plant's efficiency. So, when they could get away with not using it, they did not use it. And we woke up every morning to ash and soot.

Years later this same power company made national news at another power station the next county over. In 2014 a containment pond broke, and 39,000 tons of coal ash and 27 million gallons of wastewater flowed out of the pond and into the Dan River.[2] I grew up swimming in the Dan River; I learned to skip rocks there, at a roadside river park that had once been a foundry for the Confederate army. The coal ash that flowed into the river traveled at least seventy miles downstream. I say "at least" because of those 39,000 tons of coal ash that went into the river, only about 10 percent ever got cleaned up. Five years on, 90 percent of it is still in there, somewhere, there or downstream.[3] Coal ash is rich in all kinds of pollutants: selenium, boron, cadmium, copper, lead. After the spill, the Dan River tested especially high, dangerously high, for arsenic, barium, aluminum, and iron. It was below the thresholds for a lot of other things, but the levels were still higher than they usually were. The power company and the governments are, five years later, declaring victory and paying and collecting fines and moving on. But 90 percent of that coal ash is still in the river.[4]

Racist and Capitalist Solidarities

Oppressive and exploitative systems rely on unspoken and hidden networks of solidarity as the source of their power.[5] As a person of some privilege because of my race, gender, sexuality, class, and education, I am familiar with the sensation of being recruited into relationships of solidarity with oppressive systems. This solidarity is a bargain and a promise: accommodate yourself to this system of exploitation and domination, and you will be guaranteed a special place within it. Of course, the definition of privilege is that you get the benefits of this bargain whether you accommodate yourself or not; but the bargain itself is seductive: say yes to this system, and you won't bear the consequences of its violence. In this kind of bargain, a great many terrible actions and attitudes are permissible, and even desirable, so long as the consequences of those actions and attitudes are visited upon someone else. In the discourse of race, we see that for many people racist structures are permissible or even desirable because they affect people who are not "like me." In the discourse of capitalism we speak of externalized costs: costs not borne by either a corporation or its customers, but rather borne by the "commons" or the society or world at large, or by people so far away and so un-"like me" that they become nothing but an abstraction.

Thinking in hindsight about my small hometown, it is easy to see both racist and capitalistic solidarities at work. A deeply structural racial apartheid system governed life in that town, managing the behaviors and attitudes of everyone who lived there, and constraining actions in domains as wide ranging as work, religion, sex, commerce, discourse, recreation, and politics.[6] The logic of race was not always completely visible, but it was always completely present. Likewise, the logic of capitalism imposed and demanded its own solidarities. Tobacco dominated the economy—tobacco being an industry with incalculable externalized costs—and prosperity depended on the continued extraction of the broad green leaves from the land every fall. Every September, a number of my classmates received excused absences from school to work in their families' tobacco fields, because the demands of tobacco production surpassed the demands of education. I remember the sharpest conflict in my childhood church was over the question of smoking inside the church, because for some any attempt to limit smoking, even in sacred space, was a violation of tacit solidarity with the tobacco industry.

The coal-fired power plant I could see from the backyard depended on both kinds of solidarities, both capitalist and racist. It courted solidarity from those of us who lived nearby, reckoning the economic benefit of job opportunities would be enough to make us turn away from the externalized costs of coal ash and soot every morning. But the fact the plant existed there at all was evidence of another bargain made with more powerful people who lived somewhere else: the plant was in our backyard, and not in someone else's, precisely because under the relentless logic of capitalism, it was cheaper to put it in our town, because we were less able to say no. We were someone else's abstract others, externalized costs in the bargain they had made. And within our sharply segregated small town, with white people and

Black people clustered in neighborhoods which hardly touched and never overlapped, all of us who are familiar with the logic of race can already predict which particular part of town the Belews Creek Steam Station stood closest to.[7] It was *visible* from my backyard, but it *loomed* over others' backyards—the backyards of the town's African-American residents. Our town and region had received this power station as part of a bargain made by someone more powerful than us, but even within our town, the burden was not equally shared, and the burden was distributed on the basis of a racist logic.

Trouble in Eden (and Little Egypt)

In the summer of 2019, I sat down to lunch with two women from my hometown: Tracey and Leslie.[8] We spent ten minutes, in the way of Southerners, determining whether and how we were related, and whom we knew in common. Although both of the women claimed multiracial ancestry, by the racial logic of our hometown, one of them was clearly categorized as white, and one of them was clearly categorized as Black. Both are active as organizers against coal ash and as leaders in religious communities.

Tracey began to tell me about a community called Little Egypt. When they built the plant, the power company bought out all the families with land and homes where the creek would be dammed to create the lake which would serve the plant. "They all moved out," she said, "but they just moved to the outskirts. But they would have moved further if they had known." The anecdotes rolled in. People with stomach cancer. People with neurological problems. Young people. She mentioned another community, a larger one to the east, called Eden, where people who ate fish from the power station's lake had begun to fall ill and die. Nobody had told them not to eat the fish, she said, and so they ate it. They had planted gardens, they had eaten fruit off the trees for years, washing off that ever-present thin film of black soot. She told the story of a family, two people who worked at the plant for years, who ended up in wheelchairs with neurological disorders. Another member of the family went to the doctor, got diagnosed with cancer, and lasted two weeks. Tracey brought up story after story, and Leslie interrupted from time to time to remind her of someone she had forgotten. These were anecdotes, not data that was measurable, at least not measurable at this lunch table. But the narrative of loss was palpable. In small communities of dozens or hundreds of people, on a scale the size of families, these stories were piling up. They had been piling up for decades. Surely, they wondered, it could not be coincidence these small communities of African Americans, the very African Americans who had been displaced from their land by the plant and who had moved to new land as the plant's next-door neighbors were the ones affected? Could it be coincidence they were the very ones with all these stories about getting sick?

Leslie spoke up: it wasn't just Black people getting sick. White people had been getting sick too, sharecroppers who worked land near the plant, people downwind of the twin smokestacks like my own family and neighbors. But the white people, she said, were slower to speak up. Even later, even once there was open conflict between the community and the power company about the plant, she said, it was middle-class white people who were last to join the fight, or who never joined it at all, even though they were affected. Even though they were sick, too. Though she didn't use the word, Leslie was pointing to the power of solidarity. Polluters could count on the racist solidarity of white people who accepted harm as long as it was disproportionately aimed at Black people, and corporations could count on the solidarity of middle-class people to maintain the status quo in which they enjoyed privilege. The logics of race and capitalism were functioning as intended in my small town, both recruiting accomplices into their systems of solidarity, and among the many externalized costs were that fine layer of soot every morning and a suspiciously high number of serious illnesses and early deaths.

Change

Life in a small town feels immutable. Things rarely change, and those that do change so slowly or subtly that it's hard to perceive. My life has taken me far away from that town and my visits have become less frequent. And when I do return, I see just how incremental the changes have been: a new stoplight here, a new sign for the hardware store there, the Baptist church has paved its parking lot. But in the days of internet connections and constant social networking, I have been able to watch other kinds of change play out in real time—kinds of change that I never anticipated from my childhood home, changes that would have been unthinkable when I lived there, mediated by Facebook Live and its digital kin.

Perhaps the most surreal of these mediations of change was a moment I had, sitting in my office at the Iliff School of Theology, watching people gather on the shores of Belews Lake, in sight of those familiar gray towers from the steam station.[9] People gathered there on the shore with former vice president Al Gore and with preacher-activist Rev. Dr. William J. Barber II of the Poor People's Campaign. A reporter from the *New York Times* was there, and an article about the gathering was published in the paper a few days later.[10] It is difficult to explain just how bizarre this was to watch at a distance of 1500 miles, my small town, suddenly famous on the internet. Those dignitaries were there to talk about the steam station—to point out the environmental destruction that had been wrought by the power company there and at other sites like the coal ash spill nearby. They cited scripture and they prayed. They sang hymns like "Victory Is Mine." They were there to draw attention to all those sick people that Tracey and Leslie had told me about; they were there to speak publicly about those geographies of exploitation and pollution which had been such a fact of life for our town since I was a child.

It was a proud moment for me, but more than anything else it was bewildering, because as I looked at that crowd gathered there, I realized a great deal had changed in my absence, and that the power of both of those seemingly invulnerable solidarities—the power of the solidarity of capitalism and the power of the solidarity of race—had been eroded and stripped away. Not much had changed, but beyond stoplights and parking lots, something fundamental had changed about the economic and racial logic of that town. In the same way that solidarities of race and capitalism had once relied on each other in their construction and had drawn on each other for their power, as I watched from my computer screen, the power of race and capitalism were intertwined in their diminishment. The people in that crowd were there together on the basis of a different kind of solidarity than the kind I had grown up knowing. I had of course known plenty of times growing up when people gathered in a group across racial boundaries. But those gatherings always seemed to be governed by arcane and unwritten sumptuary codes and a kind of human zoning law; even in a public space like a Christmas parade or a football game, people gathered according to race, mimicking the segregated neighborhoods that they lived in and the schools and restaurants that had been segregated legally until just a generation earlier.[11] But watching those people gathered by the lake, it seemed the power of that old racist solidarity, the grandchild of Jim Crow, had been changed, possibly weakened and damaged, revealed to be less absolute than anyone had imagined.

Better Solidarities

The press conference with Dr. Barber and Vice President Gore was one prominent moment in a much longer arc of change. Leslie and Tracey filled me in on some of the history: as concerns about the coal ash spread, and as a company began to drill exploratory wells for fracking in town, again mostly near the African American neighborhoods, people began to question those old solidarities of capitalism and race. A group formed to advocate on the issue of coal ash, and it formed across boundaries of race. Because so much of social life was organized by religion, people organized out of churches, and with the hymns of their churches, singing "God's gonna trouble the water." Another group formed to resist the fracking wells, and this group, too, stretched across racial lines. "They needed us," Tracey told me, "and we needed them. We hugged each other," she said, "we sang together, we fellowshipped together." The power company went to its old familiar playbook; at a meeting the company held for residents along a street called Plantation Drive, whose water had been tainted by the plant, the company brought in plates of sandwiches, sodas, and chips and promised to give the residents water for free. There might have been a time when corporate benevolence would have been convincing, when those old solidarities would have taken over. But a new solidarity was forming, and the story in town became that the company was trying to buy people off cheap with sandwiches.

Both the town government and the county government had been cozy with the power plant, as might be expected. People organized across party lines and across racial lines and defeated some of those people in elections. Moratoriums were put in place for future fracking wells, moratoriums being the best they could do under a state law that prohibited local control over such things.[12] The power company came to town to test the water for pollutants from coal ash. The town's water was over the limits. But the governor of the state at the time was a former twenty-nine-year employee of the power company, and the state increased the limits so that the town's pollutants were within ranges of acceptability.[13] And so things have proceeded—the power plant's towers still jutting up over the trees, people still getting sick in mysterious clusters. But there is a sense something has shifted. In these new coalitions, which formed to resist fracking and hold the power company accountable, the old solidarities were diminished. The old bargains no longer seemed like such a good deal. The tobacco industry is very nearly gone now, something unimaginable twenty years ago, and people have discovered life has continued. People who a generation ago would not have associated with each other in public now organize and petition together, and they also eat and date and live together. It is not all because of the coal ash fight, of course, but it is hard to escape the sense that some of it is.

Praying for the Water

Last spring a Facebook video caught my eye, and it is one of a few like it that I have seen recently. In it, a small group of people, about half white and half Black, gather at that same roadside park by the Dan River where once stood a Confederate foundry, where I once swam and skipped rocks as a child. It was a prayer service for the river, which was still silted with coal ash from the spill five years before. The group prayed over the water—they prayed across theological traditions and across the strong racial boundaries that characterize denominational traditions in American Christianity. They prayed for clean water. They prayed for good health. They prayed for financial reparations and health care for people who had been affected. The prayers were punctuated by cries of "yes" and "thank you, Lord." And then a container of the water prayed for was carried to the riverside, and one of the ministers poured it symbolically back into the river. It did not mean the river was clean or that people had been restored to health, but it did signal, at least to me, something had shifted in the solidarities that governed life in that place. As Tracey said to me later as we remembered the prayer service together, what we need to do is protect people and the environment first, and then everything else can come afterward.

Notes

1. For information on this plant provided by its owner and operator, Duke Energy, see https://www.duke-energy.com/our-company/about-us/power-plants/belews-creek-steam-station.

2. See factsheet by the US Department of the Interior, "Dan River Coal Ash Spill" (2019), https://www.cerc.usgs.gov/orda_docs/CaseDetails?ID=984. A timeline of the spill and its aftermath can be found here: https://www.charlotteobserver.com/news/local/article10787168.html.

3. Catherine E. Shoichet, "Spill Spews Tons of Coal Ash into North Carolina River," *CNN*, February 9, 2014; "The Dan River Coal Ash Spill, One Year Later," *Southern Environmental Law Center*, February 2, 2015.

4. Jennifer Fernandez, "Cleaning Up and Moving Forward, Five Years after Dan River Spill," *Greensboro News & Record*, February 2, 2019.

5. My thinking about race and racism as a system of solidarity is informed by Karen E. Fields and Barbara J. Fields, *Racecraft: The Soul of Inequality in American Life* (London:Verso, 2012). I have also been strongly influenced to think of race as a system of oppression and domination that recruits persons into positions of dominator and oppressor by the work of scholars like Ta-Nehisi Coates, *Between the World and Me* (New York: Spiegel & Grau, 2015); Michelle Alexander, *The New Jim Crow: Mass Incarceration in the Age of Colorblindness* (New York: The New Press, 2012); Miguel A. De La Torre, *The Politics of Jesús: A Hispanic Political Theology* (New York: Rowman & Littlefield, 2015); Bryan Stevenson, *Just Mercy: A Story of Justice and Redemption* (New York: Spiegel & Grau, 2014); and Richard Delgado and Jean Stefancic, *Critical Race Theory: An Introduction* (New York: New York University Press, 2012).

6. One of the women I interviewed for this paper (introduced below), who identifies as African American, would nuance this statement. She recounts the friendliness of a local business owner who welcomed her and her family in his shop, despite his being the head of the Ku Klux Klan in the area. I can recall my own instances of this kind of doublespeak, or of the distance between ideology and lived experience, during my childhood in town. Living with this kind of contradiction and paradox is characteristic of life under the complex and twisted systems of solidarity produced by the practices of race and racism.

7. In this paper I use the terms "Black" and "white," despite the difficulty with those words, to represent and reproduce the vernacular of the community about which I am writing. In some circumstances I use the designation "African American," usually when I am taking the posture of an outside observer, in order to affect a more clinical stance. In all cases "Black" and "African American" are meant to be synonyms, despite the profound historical and definitional problems with that equation.

8. In the course of this conversation, both women agreed their comments were on the record and that I could use their names. They are both activists in religious and environmental activities in their own right, and both have a strong history of organizing and leadership. Leslie noted the significance of the location where we were meeting for our lunch and conversation. In 1968, the restaurant (then under different ownership) was the first one in town to be integrated. A group of African American

students, perhaps inspired by the sit-ins at lunch counters in nearby Greensboro, forced the integration of the restaurant.

9. A record of a live stream of the event can be found at: https://www.facebook.com/RevDrBarber/videos/belews-creek-press-conference-nc-ecological-justice-tour-north-carolinians-affec/2317712284920777/

10. Kendra Pierre-Louis, "A Leader in the War on Poverty Opens a New Front: Pollution," *New York Times,* August 24, 2018.

11. For a rich discussion of sumptuary codes in the practice of race, see Fields and Fields, *Racecraft,* 25–74.

12. Bertrand M. Gutiérrez, "Stokes Votes for Fracking Moratorium," *Winston-Salem Journal,* September 28, 2015.

13. Some may dispute the framing of this claim. However, it is clear that shifting modes of regulation, especially under the McCrory governorship where regulation was rolled back significantly, have sparked concern and action from citizens of the state. See Alana Semuels, "The Saga of North Carolina's Contaminated Water," *The Atlantic,* April 18, 2017. More recently, something similar has happened with federal oversight of coal ash, as the Environmental Protection Agency recommends rolling back protections. Lisa Friedman, "E.P.A. Weakens Rules Governing Toxic Water Pollution from Coal Plants." *New York Times,* November 4, 2019.

CHAPTER NINE

The Flint Water Crisis—Modernity Realized

Oscar Eduardo de Paz

The Flint Water Crisis (FWC) is the most visible incident of environmental injustice in recent memory. Public officials and researchers have used a familiar set of arguments for making sense of this crisis. Such arguments often include accident or disaster, technical failure, administrative malfeasance, and environmental racism. These explanations, however, are problematic on two counts. First, the explanations do not consider the implications of the complex interplay between humanity's faith in its technological capacity and intellectually laboring under a limited set of epistemological and ontological constraints. Second, the explanations only consider immediate factors. These explanations locate injustice's explanatory factors or key events in the near-term, ignoring significant factors or events that precede more immediate factors. As a consequence, past actions appearing unrelated since time has passed and contributing to environmental injustice occurring presently are overlooked. This short view of history ignores any reciprocal relationship between previous social and environmental actions and present socioenvironmental conditions and ramifications. Instead, human-environmental interaction is perceived in a snapshot. While immediate factors may have been the sources of the immediate occurrence of the Flint Water Crisis, it ignores the larger social, historical, and environmental context in which it takes place; the treatment and use of the Flint River historically, and the impact of cumulative injustice endured by Flint's population of racial and ethnic minorities and economically disadvantaged.[1] Historical conditions, under which a catastrophe of environmental injustice such as Flint could only occur, gave public officials and administrators little room for error.

Short-term explanations characterize environmental injustices as deviations from the proper functioning of the social order. The explanation of environmental racism presumes the same about occurrences of environmental injustice. Environmental racism implicates the sociopolitical structures and processes of modernity, but it offers no critique of its social values, principles for creating knowledge, and ideas of society's relationship to nature—the elements that social structures are built on and in which processes are perceived. Inevitably, characterizing environmental injustice as deviations from the normal social order results in attempts at structural reform; thus, reformers attempt to secure equity through the courts of law, the crafting of policy, and executive agencies' administrative actions. But what if the problem is not an ill-optimized social order but that social order's very social values, principles for creating knowledge, and ideas of society's relationship to nature? These explanations and reformist strategies give no consideration to modernity's ethos and logic, and any implications as culpable for environmental injustice. Modernity's paradigm escapes culpability. I argue that environmental injustice is not a deviation from the normal social order but the normal outcome of modernity's ethos and logic. The environmental injustice that befell Flint residents and continues to persist is the normal outcome of modernity's paradigm.

Modernity's Paradigm

Thomas Kuhn used the concept of a paradigm to explain the development of scientific thinking. He defined it as the accepted model of scientific inquiry.[2] Paradigm has moved beyond Kuhn's initial use, however; a broader use of the term refers to the assumptions, concepts, values, and practices shared by a community, constituting its way of viewing reality. In terms of a case study, a paradigmatic case illustrates a society's more general characteristics.[3] The modernist paradigm's ethos and logic are technological.[4] Modernity, as process of social development, seeks to optimize the organization and application of circumscribed scientific practice to human affairs.[5] Modernity in this sense unites the industrial era and information age to the present. Modernity, while coinciding with distinct time periods, is distinct from the periods themselves as a stand-alone set of social values, principles for creating knowledge, and ideas of society's relationship to nature.

Elements of Modernity's Technological Ethos and Logic

Modernity promises prosperity. Technological prosperity is the promise of increasing individual and societal intellectual and material improvement by maximizing individual and societal resource advantage using a circumscribed science. Future innovations for solving any potential threats or disruptions to technological prosperity are expected. Technological optimism is the belief that social, economic, and environmental disruptions resulting from the application of technological science and its artifacts can be resolved through further applications of technological science. Technological science is a mode of inquiry that is conducted under a set of strict principles for constructing knowledge and assumptions about nature-society relations to produce instrumentation designed for social and environmental control. It is the means for achieving technological prosperity. Its success is the reason for modernity's optimism. Technological optimization, the continuing refinement of this particular social order and its problem-solving capacities, according to the limited set of epistemological, ontological, and axiological criteria of technological science, completes modernity's paradigm; it is the consistent application of technological science to all of society.

The outcome of the paradigm is the regular occurrence of social and environmental injustice against racial and ethnic minorities and the economically disadvantaged, which are widespread global phenomena.[6] The regular, systemic occurrence of environmental injustice from modernity amounts to a type of racism. Since modernity's racism—the disproportionate burdening of adverse social, economic, political, and environmental conditions on vulnerable populations—is attributable to the prevailing paradigm, it is referred to as paradigmatic racism. This concept implies that modernity's dominant set of epistemological, ontological, and axiological constituents are the bedrock of racist praxis and outcomes.

The Commodification of the Flint River

Modernity's promise of technological prosperity requires repurposing nature from its ecological purpose and transforming it into commodities in order to satisfy our drive for individual or societal resource advantage; thus, nature's value is understood as deriving from its use in pursuing technological prosperity. The Flint River's commodification began as early as the 1830s with the development of Flint's first industries—lumber mills, paper mills, and chemical processing. The carriage and automobile industries' development continued the process. Flint's earliest industry, lumber mills, began the pattern of dumping untreated waste into the river; this pattern continued with the newer industries of paper mills and chemical processing. The historical misuse of the Flint River as a dumping ground for industrial waste coincided with its commodification. During Flint's industrial zenith, the Flint River and its tributaries were used as a drain by eight divisions of General Motors, which dumped large quantities of hazardous chemical waste directly into it.[7] Additionally, the Vogt Packing Company, a rendering plant, with permission from the local government, dumped waste into the river.[8] As a US Department of Interior report states, the wastes included chemicals such as cyanide, hexavalent chromium, nickel, oil, and suspended solids.[9] The dumping of pollutants into the river continued unabated, while the city also used the river as its primary water source until it switched sources in 1967.[10] The continued contamination was noted in a report conducted seven years later by Michigan's Department of Natural Resources.[11]

Modernity notes humanity's impact on the processes of the natural world; however, the social and natural architectures are conceived as being maintained by distinct relations and rules.[12] The dichotomy of nature and society thinks of them as functioning similarly but as phenomenal structures standing apart.[13] This dichotomy is characteristic of early modernity's view of society and nature's relation to one another. It is also visible in current arguments skeptical of society's ability to alter the earth's climate or other natural processes.[14] The duality renders parts of nature (e.g., rivers, atmosphere) as inexhaustible dumping grounds for pollution. Commodification requires nature to be understood as both an endless source of resources and a repository of waste from processes of production.[15] In pursuing technological prosperity, natural resources are used excessively, depleted, and degraded. Even in its modest form, recognizing limits and the efficient use of natural resources, modernity assumes the commodification of nature for technological prosperity and the power of technological innovation to overcome natural limits.[16]

In the context of modernity, growth is not just an expectation or prediction, but an imperative for inducing technological prosperity. In other words, it is not a rational conclusion drawn from the available evidence alone, but a social virtue of imperative importance to be pursued. While it is commonly held that the pursuit of societal resource advantage is actually good for maintaining the environment, it is evident to many that resource depletion and pollution result from the unfettered pursuit of technological prosperity, with negative effects disproportionally afflicting vulnerable populations.[17] Vulnerable populations find themselves in an even more precarious position from the seizing and overuse of natural resources since often their survival depends on them. The conditions of poverty for many populations whose quality of life and very survival is dependent on the safe access to local natural resources across the globe are fabricated by actors independent of the local ecology seizing the natural resources of a locality and subsequently extracting the raw resources, leaving a depleted environment and waste.[18] While the literature on environmental injustice notes this phenomenon occurring in the Global South, Flint provides a fresh reminder of early modernity's legacy in the Global North. The extraction of natural resources (e.g., oil, mining) has a long historical record of creating conditions of poverty for vulnerable local populations.[19] In addition to the production of poverty, extractive industries disempower local communities politically.[20]

The government facilitates industry's commodification of nature. Fearful that economic growth will slow, government's acceptance of technological prosperity as a political imperative prevents it from challenging environmentally harmful industrial practices.[21] Instead, public officials use their power to advance the cause of technological prosperity and the interests of industry over constituents' interests by preventing regulation.[22] Governments enter contracts for natural resources that favor industry over local communities in trying to achieve technological prosperity and resource advantage.[23] In some cases, the government abdicates its power and management of the stock of natural resources to private entities; thus, resource development hinges on the practices and policies of private industries.[24] In other cases, the government, in its ambition to grow an industrial base and increase economic growth, uses its power and authority to appropriate decision-making power from the public and obscure its agenda from public scrutiny.[25] The creation of adverse environmental conditions for vulnerable populations as a result of nature's commodification and industry and governments' pursuits of increasing resource advantage traverses the globe.[26]

As a commodity, the Flint River's degradation severely diminished its value as a natural resource. The river's value was dependent on its significance as a resource advantage or in contributing to attaining technological prosperity. No longer considered a valued resource to industry, public officials and business leaders sought alternatives to using the Flint River as a primary source of water. The argument for the switch claimed the river was unable to meet the demand from industry and residents. Public officials, industry, and research consultants expected Flint to continue growing.[27] A contracted engineering firm suggested two options to Flint: contract with Detroit or build direct access to Lake Huron.[28] Initially, Flint's City Commission voted in favor of Flint building direct access to Lake Huron. However, after local business executives and institutions such as the Chamber of Commerce placed their support behind purchasing from Detroit to meet the increasing industrial and population demand expected, the City Commission rescinded its previous decision and voted in favor of purchasing from Detroit, despite opposition from residents.[29] Nonetheless, the result had the same effect: resource reach.[30]

The Reemergence of the Flint River as a Valuable Commodity

The Flint River's value as a potential commodity remerged after years of failed economic development schemes by Flint.[31] Its reemergence as a valuable commodity was predicated on two conditions; its significance in either representing a resource advantage or pursuing technological prosperity and the technological optimism of area officials and leaders. A very early proposal for ensuring Flint could meet the desired industrial and population expansion demanded was to build a pipeline from Lake Huron; it was seen as an option that could provide for Flint alone or Genesee County.[32] Fifty years later, a similar proposal, now championed by Genesee County officials, was proclaimed a viable option for returning technological prosperity to Flint and the county. Modernity's vision of technological prosperity is not fleeting. Several arguments were advanced, all along a line of increasing the resource advantage of individuals or society. State, county, and city officials touted that building the pipeline to Lake Huron would result in savings for the city and its residents.[33] The various emergency managers strongly promoted the cost savings for Flint and its residents.[34] County officials particularly pressed the argument that increased water access would draw agro-businesses, spurring economic development in the region.[35]

Flint and Genesee County officials pushed a common theme, resource independence, specifically, water independence along with common economic development arguments.[36] Technological prosperity promotes competition among political actors for resources in national and regional settings. This competition is regularly framed as resource independence to justify parochial actions taken in securing energy resources. The concept of resource independence entails sovereignty; a conceptual and political idea that suggests separate political identities and fates, sovereignty is an assertion of independence from other political actors.[37] Resource independence undermines regional cooperation and coordination.[38] It is in tension with cooperative governance among political actors.[39] Robust claims of sovereignty encourages political actors to ignore other actors' interests.[40] Moreover, political actors claiming resource independence ignore how their policies effect other actors.[41] We can see this play out in Flint. In April 2013, Flint's emergency manager notified the state of the city's decision to switch to the Karegnondi Water Authority (KWA); the following day, the Detroit Water and Sewerage Department notified Flint that its service would terminate in a year, April 2014.[42] The abrupt cutoff as a result of the resource competition flare-up between the two cities created a time constraint that left Flint with little choice but to rush in preparing its water infrastructure and personnel for the switch, which would have tragic consequences down the line.

Technological Optimism—Dismissing Challenges and Ignoring Warning Signs

The Flint River's recommodification as a water source despite a well-known history of toxicity and degradation was possible because of the permeation of modernity's paradigm in the thought and practice of politics, policy, and administrative decision-making. Technological optimism empowers decision-makers to make decisions on the basis that current technological solutions and future technological advances will be available to remedy past errors and avert or limit the impact of potential problems, ignoring valid concerns even when raised by its own technological science. The City of Flint spent nearly four million dollars upgrading its water treatment plant to treat water from the Flint River.[43] It was known that the treatment of the Flint River presented greater challenges than a deep lake source such as Lake Huron.[44] The river contained greater levels of organic matter, requiring treatment with chlorine; however, chlorine and organic material can interact and form total trihalomethanes (TTHM)—a known carcinogen. In January 2015, as a result of elevated levels of TTHM, the City of Flint was cited for violating the Safe Drinking Water Act.[45] This only becomes a problem after increasing the level of chlorine to treat the water in response to tests indicating elevated levels of E. coli and coliform bacteria, leading the government officials to advise residents to boil their water.[46] Flint's dilemma illustrates a unique problem in modernity's relationship with technology and science. In modernity, technological science is not only our means for resolving and protecting society from environmental threats or crises but is also a potential source of the threats and crises.[47] Indeed, early communications between public officials indicate that a significant problem could develop after the switch. Flint's water quality supervisor advised against distributing the river's water anytime soon after the April 2014 switch.[48]Another public official expressed a warning that the rapid move to the Flint River prevented the recruitment of bids from qualified vendors to complete the necessary infrastructure improvements to Flint's water treatment facility, potentially resulting in disasters.[49] This circumstance resulted from the resource competition stoked by pursuing technological prosperity. Nevertheless, state, county, and city officials gathered at the water treatment plant celebrating the switch to the Flint River with a celebratory glass of water.[50]

Choosing Technological Prosperity over Democratic Governance

The political disempowerment of Flint residents preceded the Flint River's recommodification. Not at all surprising to find, since extractive industries disempower local communities politically.[51] Already noted earlier, government, more often than not, sides with the needs and demands of industry and technologically driven prosperity over the concerns of the public or at the very least assumes the public's concerns neatly align with the pursuit of technological prosperity—an assumption that environmental conflicts around natural resource extraction demonstrates as patently false on a global scale.[52] Indeed, very often, the values of the public, industry, and public officials not only do not align but are incommensurable.[53] Many of the key decisions that cascaded to become known as the Flint Water Crisis occurred under the watchful eye of the state's emergency management of Flint. The state justified emergency management by declaring Flint in a state of financial emergency—a determination reached by an unelected, government-appointed financial review team of eight members.[54] On two occasions, the then governor expanded the powers of emergency managers, only to have, in one case, the public, through a ballot proposal, repeal the law two years later.[55] Nevertheless, in Flint, over a nearly ten-year period under emergency management, emergency managers used directives and executive orders to make decisions charting Flint's economic, political, and environmental future absent any mechanism of accountability to the public, such as elections or administrative oversight.[56] Early on under emergency management, the offices of the city's ombudsman were shuttered, effectively further undermining accountability to the public.[57] While the decision to switch water sources from the Detroit Water and Sewerage Department was approved by city council members, using the Flint River as an interim water source was strictly decided upon by the then acting emergency manager in consultation with other public officials.[58] The technological prosperity imperative and the drive for resource advantage prohibits putting up with practices that inhibit the pursuit, even fair, democratic governance. Emergency management privileged the pursuit of technological prosperity, weakening public participation and oversight in order to accomplish its goal.

Technological Science—Burdening the Community

The Flint Water Crisis, similar to other environmental injustice crises, occurs in the purview of a scientific practice that labors under limited principles for constructing knowledge and conceptions of nature-society relations. Technological science refers to a mode of inquiry that is conducted under a set of strict epistemological and ontological assumptions to produce instrumentation designed for social and environmental control. It is the means for achieving technological prosperity. Modernity's orientation toward a progressive future places great significance on questions relevant to determining rational means of knowing and how society acquires and increases intellectual and material abundance and prosperity, satisfying the drive for a resource advantage. In broad strokes, the science discussed has become commonly labeled as positivism. The term positivism is avoided; instead, technological science is preferred because the term is inclusive of the characteristics of positivism but does not lose sight of its relation to the other characteristics of modernity. The commitment to labor under a single set of ontological and epistemological constraints is a uniquely defining characteristic of the science that informs modernity; the commitment is based on the belief that technological science results in the singularly, nearly absolute correct answer to an object of inquiry.[59]

Technological science becomes a barrier to communities attempting to tackle environmental injustice. Technological science's focus on causality—that is, the separation of exogenous factors in order to isolate the principle variable exerting influence—sets a standard that is difficult to achieve with a complex social phenomenon such as environmental injustice.[60] In the case of Flint, the validity of scientific evidence became a source of conflict. In the early stages of the Flint Water Crisis, the Michigan Department of Health and Human Service's scientific experts attributed the elevated levels of childhood blood levels to expected seasonal changes.[61] This was contentious because the Hurley Medical Center in Flint found childhood blood levels increased compared to results from an earlier test when Flint received its water from Lake Huron.[62] It would take months after the start of the FWC for an agreement in findings to occur, and even then, public officials used the criterion of causality to avoid attributing any responsibility to the water source switch.[63] The discussion between public officials and government and nongovernment scientific experts concentrated on scientific techniques, questions of appropriate measurement and scientific procedure, taking precedence over risk to the public and potential harm.[64]

Public officials use technological science and its standard of causality as the measure of communities' claims, while loosely applying science and a weaker standard of causality to their own decisions with environmental justice implications.[65] This seems most common in the economic arguments used by public officials claiming that projects will deliver jobs and technological prosperity. In this context, other means of knowing are dismissed and disparaged. Nearly immediately after the city began to use the river as its water source, residents complained of foul smelling, discolored water flowing from their faucets.[66] Public officials dismissed residents' complaints about the water's aesthetic qualities with charges that while unpleasant, the odor and discoloration are part of a normal process, and the water was still safe for consumption.[67] Residents also complained of skin irritation and hair loss, believing it was related to the water source switch.[68] Within a technological science frame of reference, the possibility of other ways of knowing is nearly inconceivable.[69] Claims from aesthetic and personal experience are derided and dismissed by characterizing them as anecdotal, even though the context and implication could be deadly. Initially, residents' claims were met with skepticism; more recent research has suggested that large variations in chlorine levels, pH, and hardness could have contributed to the residents' rashes and other skin irritations in Flint; however, researchers avoided any definitive claim because of the lack of historical data on Flint's water condition.[70] Public officials dismissed concerns as attempts to use the issue as a political football.[71]

Unlike government and industry, local communities rarely have access to professionals with scientific training, needing to go outside the community to find professional help to advance their cause.[72] This condition was on significant display in the Flint Water Crisis. Recognition did not occur until highly trained professionals, one prominently from outside of the Flint region, were alerted to the potential crisis. Using their technical background and credentials, they were able to build a case demonstrating the extent of the problem and the most likely source.[73] The lack of access to scientifically trained professionals is not the only access restriction faced by communities and organizations. The circulation of academic research relevant to environmental injustice "is limited to professionals within academia."[74] Thus, communities are usually at a considerable disadvantage when it comes to the use of science and the standard of causality in justifying any claims of environmental injustice.

Technological Optimization—Keeping Up the System

When confronted with a crisis in the matrix of technology, society, and environment, powerful stakeholders such as public officials and administrators only advance solutions that maintain or regulate the previous social architecture or attempt to optimize it. Within the epistemological and ontological limits of technological science, problems are only conceived in technical terms. Technological science makes any analysis of the problem beyond the strictly defined epistemological and ontological limits nearly impossible. The constraints on imagination and insight from professional training is significant in environmental injustice's persistence.[75] The dominate form of scientific practice demands narrow specializations, which has helped it achieve success, but it has particular drawbacks.[76] Scientific expertise reduces the range of possible solutions that professionals/experts are trained in and, thus, equipped to consider; how solutions' fit in the broader context is overlooked.[77] With the Flint crisis bourgeoning, the range of solutions Michigan officials considered primarily concentrated on re-establishing some sense of order to socio-environmental-technology relations. A highly heralded and much-publicized solution was the distribution of free water filters to residents. In internal communications, concerned with public perception, public officials gloated that the program "went over extremely well with residents," that there was a demand for more filters, and that this action demonstrated officials', commitment to action.[78] The Michigan State government's response to the FWC primarily concentrated on diagnostic testing for residents, technological correctives to the water system, and increased participation of scientific experts.[79] This was suggested, even though, as already noted, the government usually houses vast troves of scientific expertise, especially in comparison to local communities.

Discussion

The Flint Water Crisis demonstrates that cases of environmental injustice are not deviations from the normal social order but a normal outcome of modernity's social values, principles for creating knowledge, and ideas about the relationship between nature and society: what has been referred to in this essay as modernity's paradigm. The paradigm was broken down into four concepts that articulate its values, principles, and ideas on nature and society's relationship—technological prosperity, technological optimism, technological science, and technological optimization.

To pursue technological prosperity society must adopt a specific orientation toward nature. Nature is repurposed from its ecological function and commodified to serve the narrow purpose of signifying a resource advantage or as an input for achieving technological prosperity. The Flint River was used as an inexhaustible source of water and a bottomless sewer for the city and industry. Since modernity's promise of technological prosperity is not an inevitable nor linear process of development but includes social, economic, and environmental disruptions, belief in technological optimism is necessary to ignore the clear and present threat that disruptions such as the degradation of the river pose to local populations and any plans to revive technological prosperity. The premise of a dichotomy of nature and society dilutes the magnitude of any potential threats polluted nature may appear to pose.

While technological science appears to grant society the means for controlling nature and society to their benefit, it aids in contributing to environmental injustice by stripping away the legitimacy of the public's informal or lay analyses of problems. Lay means of intimating a problem, such as aesthetic and personal experience, are discredited, creating a situation in which environmental problems posing potential harm are only discernable through advanced methods of detection and analysis not readily accessible by vulnerable communities. This is made even more problematic because of an incongruence in the time it takes to conduct advanced detection and analyses and the immediacy of any risk for harm to a community. As a result, public officials are slow to act. Furthermore, without a precautionary principle, that is, not acting when uncertain about the possibility of substantial harm, remedies are usually proposed after residents have been exposed to risk or suffered harm. When action is taken, public officials and relevant professionals propose solutions that support the status quo, fixating on the optimal technology fix; thus, transformational solutions are not considered or explored. Reconceptualization of societal relations toward nature are given no consideration at all. Public officials do not propose organizing a committee to study the development of an entirely new system of water distribution not readily found in existing models of privatization or government service provision. The most transformational solutions discussed (e.g., replacing the entire network of water pipes) usually still seek to maintain the existing sociotechnical system and its prevailing relations.

The outcome of modernity's paradigm is the disproportionate

burdening of racial and ethnic minority groups and the economically disadvantaged, groups that are nearly indistinguishable within the paradigm, with unjust environmental conditions. The basis of this pattern is grounded in modernity's social values, principles for creating knowledge, and ideas of society's relationship to nature. This inequitable distribution of negative outcomes from the paradigm amounts to a type of racism—paradigmatic racism. The importance of distinguishing the source of the problem is tied to the development of solutions. As noted earlier, explanations focusing on the structure of the social order propose solutions intended to fix the structural features of that social order. However, if the structural features are not the principal problem, then likely the problems will persist. Correctly determining the problem's source allows for solutions that address the issue—modernity's paradigm.

Notes

1. "Cumulative injustice" speaks to the cumulative impact from the multiple forms of injustice suffered by racial and ethnic minorities. See Andrew R. Highsmith, "Demolition Means Progress: Urban Renewal, Local Politics, and State-Sanctioned Ghetto Formation in Flint, Michigan" *Journal of Urban History* 35, no. 3 (March 2009): 348–68.

2. Thomas S. Kuhn, *The Structure of Scientific Revolutions* (Chicago: University of Chicago Press, 2012), 33–37.

3. Bent Flyvbjerg, *Making Social Science Matter: Why Social Inquiry Fails and How It Can Succeed Again* (Cambridge: Cambridge University Press, 2001), 79–81.

4. Technology is understood as the entire societal apparatus applying a circumscribed science to human affairs. Technology includes the institutions and social groups responsible for the application and implementation of scientific knowledge to human affairs. Additionally, it references the intellectual artifacts and means for constructing further knowledge (e.g., models) for giving corporeality to ideas. As defined, technology is not apolitical. As a social process, technology shapes social relations; it reflects dominate thoughts and behaviors; it is used to exert power over people and nature (e.g., Ellul, Marcuse, Redclift, Winner, Mumford).

5. Saint-Simon is considered by many the original visionary of the sociotechnical society that is at the heart of what this paper calls modernity. See Krishan Kumar, *Prophecy and Progress: The Sociology of Industrial and Post-Industrial Society* (New York: Penguin, 1978), 30.

6. Leah Temper, Federico Demaria, Arnim Scheidel, Daniela Del Bene, and Joan Martinez-Alier, "The Global Environmental Justice Atlas (EJAtlas): Ecological Distribution Conflicts as Forces for Sustainability" *Sustainability Science* 13, no. 3 (May 2018): 573–84.

7. Detroit Program Office, *Report on Water Pollution in the Lake Huron Basin: Flint River* (Grosse Ile, MI: US Department of the Interior, Federal Water Pollution Control Administration, December 1966), 27–28.

8. Ibid., 40, 43.

9. Ibid., 28.

10. W. Osmund Kelly, Tom Kay, and W. Dale Bachtel, "Flint's Industrial Water Supply—Joint Discussion" *American Water Works Association* 63, no. 3 (March 1971): 148–53.

11. Environmental Protection Bureau, "Flint River Study, August 6–7, 1974" (Lansing: Michigan Department of Natural Resources, February 1977), 5–7.

12. John Byrne, Leigh Glover, and Cecilia Martinez, "The Production of Unequal Nature," in *Environmental Justice: Discourses in International Political Economy,* ed. John Byrne, Leigh Glover, and Cecilia Martinez (New York: Routledge, 2017), 261–91.

13. Ibid.

14. Gregg Easterbrook, *A Moment on the Earth: The Coming Age of Environmental Optimism* (New York: Viking, 1995), 257–67.

15. Byrne, Glover, and Martinez, "The Production of Unequal Nature," 261–91.

16. United Nations World Commission on Environment and Development, *Our Common Future,* ed. Volker Hauff (Oxford: Oxford University Press, 1987), 156.

17. Robert D. Bullard, *Dumping in Dixie: Race, Class, and Environmental Quality,* 2nd ed. (Boulder, CO: Westview, 1994), 9–11; Ramachandra Guha and Joan Martinez-Alier, *Varieties of Environmentalism: Essays North and South* (London: Earthscan, 1997), 31.

18. Guha and Martinez-Alier, *Varieties of Environmentalism,* 12–16; Beverly Wright, "Race, Politics and Pollution: Environmental Justice in the Mississippi River Chemical Corridor," in *Just Sustainabilities: Development in an Unequal World,* ed. Julian Agyeman, Robert D. Bullard, and Bob Evans (New York: Earthscan, 2003), 125.

19. Ibid.

20. Guha and Martinez-Alier, *Varieties of Environmentalism,* 118–20.

21. Andrew Hurley, *Environmental Inequalities: Class, Race, and Industrial Pollution in Gary, Indiana, 1945–1980* (Chapel Hill: University of North Carolina Press, 1995), 97.

22. Hurley, *Environmental Inequalities,* 97–98; Diane Sicotte, "Power, Profit and Pollution: The Persistence of Environmental Injustice in a Company Town" *Human Ecology Review* 16, no. 2 (Winter 2009): 141–50.

23. Guha and Martinez-Alier, *Varieties of Environmentalism,* 5.

24. S. Harris Ali, "The Political Economy of Environmental Inequality: The Social Distribution of Risk as an Environmental Injustice," *Speaking for Ourselves: Environmental Justice in Canada,* ed. Julian Agyeman, Peter Cole, Randolph Haluza-DeLay, and Pat O'Riley (Vancouver: University of British Columbia Press, 2009), 99.

25. Leonard Nevarez, "Just Wait until There's a Drought: Mediating Environmental Crises for Urban Growth" *Antipode* 28, no. 3 (1996): 246–72.

26. Guha and Martinez-Alier, *Varieties of Environmentalism,* 40-45.

27. Kelly, Kay, and Bachtel, "Flint's Industrial Water Supply," 148–53.

28. Ibid.

29. Ibid.

30. "Resource reach" refers to a city's tendency to move beyond its local resource base, looking for raw resources further away to support its own technological prosperity.

31. Michael Moore's documentaries *Roger and Me* and *The Big One* provide a brief discussion of Flint's economic development aspirations.

32. Kelly, Kay, and Bachtel, "Flint's Industrial Water Supply," 148–53.

33. Dominic Adams, "Flint City Council Approves Resolution to Buy Water from Karegnondi, State Approval Still Needed," *Flint Journal,* March 26, 2013; Shaun Byron, "KWA Construction Starts Soon, Pipeline May Not Reach Genesee County until 2015" *MLive,* June 28, 2013; Dave C. "KWA Pipeline- Jeff Wright," West Flint Business Association, April 23, 2015, https://web.archive.org/web/20151020061843/http://westflintbiz.com/kwa-pipeline-jeff-wright.

34. Edward L. Kurtz, *Resolution to Purchase Capacity from Karegnondi Water Authority,* Resolution # 2013EM041 (Flint, MI: March 29, 2013).

35. Jeff Hogan, "KWA could pump more than water into our community," *County Press,* March 4, 2015, 4A–5A.

36. Theresa Lark and Jeff Hammer, "'The Future Is Now' Karegnondi Water Authority Breaks Ground," *MACDC Pipeline,* 22, no. 3 (Fall 2013): 20–22.

37. Anasua Basu Ray Chaudhury, "The Energy Crisis and the South Asian Security: An Indian Perspective," *India Quarterly: A Journal of International Affairs* 65, no. 2 (April 2009): 137–51.

38. Ibid.

39. Kacper Szulecki and Kirsten Westphal, "The Cardinal Sins of European Energy Policy: Nongovernance in an Uncertain Global Landscape," *Global Policy* 5 (October 2014): 38–51.

40. Ibid.

41. Ibid.

42. Merrit Kennedy, "Lead-Laced Water in Flint: A Step-By-Step Look at the Makings of a Crisis," *National Public Radio: The Two-Way,* April 20, 2016.

43. David Murray, email message to officials at the Governor's Office, February 1, 2015, in "Gov. Snyder Emails on Flint Water Crisis," *Flint Journal,* January 20, 2016, https://www.scribd.com/doc/296110751/Gov-Snyder-emails-on-Flint-water-crisis.

44. Ibid.

45. Kennedy, "Lead-Laced Water."

46. Ibid.

47. Byrne, Glover, and Martinez, "The Production of Unequal Nature," 261–91.

48. Michael Glasgow, email message to officials at Michigan's Department of Environmental Quality, April 17, 2014, in "Gov. Snyder Emails on Flint Water Crisis," *Flint Journal,* January 20, 2016, https://www.scribd.com/doc/296110751/Gov-Snyder-emails-on-Flint-water-crisis.

49. Brian Larkin, email message to officials in the Governor's Office, March 14, 2014, "Gov. Snyder Emails."

50. Dominic Adams, "Closing the Valve on History: Flint Cuts Water Flow from Detroit after Nearly 50 Years," *Flint Journal,* April 25, 2014.

51. Guha and Martinez-Alier, *Varieties of Environmentalism,* 40–45.

52. Joan Martinez-Alier, *The Environmentalism of the Poor,* United Nations Research Institute for Social Development (UNRISD) and University of Witwatersrand (August 30, 2002): 1–56.

53. Ibid.

54. Oona Goodin-Smith, "Flint's History of Emergency Management and How It Got to Financial Freedom," *Flint Journal,* January 16, 2018.

55. Ibid.

56. Ibid.

57. Ibid.

58. Ibid.

59. Frederick Winslow Taylor, *The Principles of Scientific Management* (New York: Harper, 1911), 25; Richard B. Norgaard, *Development Betrayed: The End of Progress and a Coevolutionary Revisioning of the Future* (New York: Routledge, 1994), 156.

60. Bunyan Bryant, "Issues and Potential Policies and Solutions for Environmental Justice: An Overview," in *Environmental Justice: Issues, Policies, and Solutions,* ed. Bunyan Bryant (Washington, DC: Island, 1995), 9.

61. Jennifer Eisner, "MDHHS Releases October Data Outlining Blood Lead Levels in Flint" (Lansing: Michigan Department of Health and Human Services, December 3, 2015).

62. Associated Press, "Doctor's Urge Flint to Stop Using Water from Flint River," *Crain's Detroit Business,* September 28, 2015.

63. Gov. Newsroom, email message to Rick Snyder, October 2, 2015, in "Gov. Snyder Emails on Flint Water Crisis," *Flint Journal,* January 20, 2016, https://www.scribd.com/doc/296110751/Gov-Snyder-emails-on-Flint-water-crisis.

64. Miguel A. Del Toral, *High Lead Levels in Flint, Michigan—Interim Report* (Washington, DC: United States Environmental Protection Agency—Region 5, June 24, 2015).

65. Bryant, "Issues and Potential Policies and Solutions for Environmental Justice," 9–10, 14.

66. Ron Fonger, "Lead Leaches into 'Very Corrosive' Flint Drinking Water, Researchers Say," *Flint Journal,* September 5, 2015.

67. David Murray, email message to officials at the Governor's Office, February 1, 2015, in "Gov. Snyder Emails on Flint Water Crisis," *Flint Journal,* January 20, 2016, https://www.scribd.com/doc/296110751/Gov-Snyder-emails-on-Flint-water-crisis.

68. Julia Lurie, "Meet the Mom Who Helped Expose Flint's Toxic Water Nightmare," *Mother Jones,* January 21, 2016; Unified Coordination Group, *Flint Rash Investigation: A Report on Findings from Case Interviews, Water Testing, and Dermatologic Screenings for Rashes That Developed or Worsened after October 16, 2015* (2016), 9.

69. Vandana Shiva, *The Violence of Green Revolution: Third World Agriculture, Ecology and Politics* (London: Zed Books, 1991), 233.

70. Unified Coordination Group, *Flint Rash Investigation,* 9.

71. Dennis Muchmore, email message to Gov. Snyder and officials at Governor's Office, September 25, 2015 in "Gov. Snyder Emails on Flint Water Crisis," *Flint Journal,* January 20, 2016, https://www.scribd.com/doc/296110751/Gov-Snyder-emails-on-Flint-water-crisis.

72. Bullard, *Dumping in Dixie,* 15; Conner Bailey, Kelly Alley, Charles E. Faupel, and Cathy Solheim, "Environmental Justice and the Professional," in *Environmental Justice: Issues, Policies, and Solutions,* ed. Bunyan Bryant (Washington, DC: Island, 1995), 38.

73. Fonger, "Lead Leaches"; Mona Hanna-Attisha, Jenny LaChance, Richard Casey Sadler, and Allison Champney Schnepp, "Elevated Blood Lead Levels in Children Associated with the Flint Drinking Water Crisis: A Spatial Analysis of Risk and Public Health Response" *American Journal of Public Health* 106, no. 2 (2016): 283–90.

74. Bailey, Alley, Faupel, and Solheim, "Environmental Justice and the Professional," 42.

75. Ibid., 37.

76. Ibid.

77. Ibid., 37; Ali, "The Political Economy of Environmental Inequality," 97.

78. Harvey Hollins, email message to Gov. Rick Snyder and officials at the Governor's Office, September 5, 2015; Nick Lyons, email message to Gov. Rick Snyder and state officials, October 6, 2015 in "Gov. Snyder Emails on Flint Water Crisis," *Flint Journal,* January 20, 2016, https://www.scribd.com/doc/296110751/Gov-Snyder-emails-on-Flint-water-crisis.

79. John Roberts, email message to Gov. Rick Snyder and officials at the Governor's Office, October 7, 2015 in "Gov. Snyder Emails."

Acknowledgments

The degradation of the environment is the greatest threat facing humanity. The survival of the human race is dependent on the health of our planet. And while all will eventually face the ramifications of ignoring the cry of the earth, those who are relegated to the underside of history, predominately those who are of color, disproportionately face the negative consequences. All too often, the current discourse concerning environmental justice ignores the voices of those most impacted. As a corrective, the Iliff School of Theology made a commitment to the conversation by focusing on marginalized, minoritized voices. To that end, a conference was held in October 2019 to discuss the intersection of spirituality and environmental racism. We acknowledge and are grateful to the Iliff School of Theology, its board of trustees, and its president, Thomas V. Wolfe, for their vision and commitment to the discourse. Thanks to their financial support in holding a conference on the impact of water on disenfranchised communities, we are able to share the presentations made during those days with you.

We also must recognize Rachel Hackenberg, editor at The Pilgrim Press. During a private social gathering in New York City, the conference on water was mentioned in passing, and Mx. Hackenberg immediately expressed interest in making the findings of the conference available to a larger audience. Thanks to her foresight, you now hold in your hands this book. Finally, I would be remiss if all who participated in the conference were not also recognized. We express our gratitude to them, those whose presentations made it into the book, and those whose did not.

Bibliography

Alexander, Michelle. *The New Jim Crow: Mass Incarceration in the Age of Colorblindness.* New York: The New Press, 2012.

Ali, S. Harris. "The Political Economy of Environmental Inequality: The Social Distribution of Risk as an Environmental Injustice." In *Speaking for Ourselves: Environmental Justice in Canada,* edited by Julian Agyeman, Peter Cole, Randolph Haluza-DeLay, and Pat O'Riley. Vancouver: University of British Columbia Press, 2009.

Bailey, Conner, Kelly Alley, Charles E. Faupel, and Cathy Solheim. "Environmental Justice and the Professional." In *Environmental Justice: Issues, Policies, and Solutions,* edited by Bunyan Bryant. Washington, DC: Island, 1995.

Barton, William. *Old Plantation Hymns: A Collection of Hitherto Unpublished Melodies of the Slave and the Freedman, with Historical and Descriptive Notes.* Boston: Lamson, Wolffe, 1899.

Bhabha, Homi. *The Location of Culture.* New York: Routledge, 1994.

Boff, Leonardo. *Cry of the Earth, Cry of the Poor.* Translated by Phillip Berryman. Maryknoll, NY: Orbis Books, 1997.

brown, adrienne maree. *Emergent Strategy: Shaping Change, Changing Worlds.* Chico, CA: AK Press, 2017.

Bullard, Robert D. *Dumping in Dixie: Race, Class, and Environmental Quality,* 2nd ed. Boulder, CO: Westview, 1994.

Bullard, Robert D., Glenn S. Johnson, and Angel O. Torres. *Environmental Health and Racial Equity in the United States: Building Environmentally Just, Sustainable, and Livable Communities.* Washington, DC: American Public Health Association, 2011.

"Business: Running Dry; Business and Water." *The Economist* 388, no. 8594 (August 23, 2008): 53–54.

Butler, Octavia. *Parable of the Sower.* New York: Grand Central Publishing, 2007 [1993].

Bryant, Bunyan. "Issues and Potential Policies and Solutions for Environmental Justice: An Overview." In *Environmental Justice: Issues, Policies, and Solutions,* edited by Bunyan Bryant. Washington, DC: Island, 1995.

Byrne, John, Leigh Glover, and Cecilia Martinez. "The Production of Unequal Nature." In *Environmental Justice: Discourses in International Political Economy,* edited by John Byrne, Leigh Glover, and Cecilia Martinez. New York: Routledge, 2017.

CERD Working Group on Health and Environmental Health. *Unequal Health Outcomes in the United States: Racial and Ethnic Disparities in Health Care Treatment and Access, the Role of Social and Environmental Determinants of Health, and the Responsibility of the State.* Report to the UN Committee on the Elimination of Racial Discrimination. New York: United Nations, 2008.

Cesar, Sylvia. "Privatization of Water: Evaluating Its Performance in the Developing World." *Annals of Public and Cooperative Economics* 90, no. 1 (July 2018): 5–23.

Chaudhury, Anasua Basu Ray. "The Energy Crisis and the South Asian Security: An Indian Perspective." *India Quarterly: A Journal of International Affairs* 65, no. 2 (April 2009): 137–51.

Choi, Hee An. *Korean Women and God: Experiencing God in a Multi-Religious Colonial Context*. Maryknoll, NY: Orbis Books, 2005.

Clare, Eli. *Brilliant Imperfection: Grappling with Cure*. Durham, NC: Duke University Press, 2017.

Clement, Andrew, Krista Boa, Simon Davis, and Gus Hosein. "Towards a National ID Card for Canada? External Drivers and Internal Complexities." In *Playing the Identity Card: Surveillance, Security, and Identification in Global Perspective*, edited by Colin J. Bennett and David Lyon. New York: Routledge, 2008.

Clinton, Catherine. *Harriet Tubman: The Road to Freedom*. New York: Little, Brown, 2004.

Coates, Ta-Nehisi. *Between the World and Me*. New York: Spiegel & Grau, 2015.

———. *The Water Dancer*. New York: OneWorld, 2019.

Connor, David J., Beth A. Ferri, and Subini A. Annamma, eds. "Dis/Ability Critical Race Studies (DisCrit): Theorizing at the Intersections of Race and Dis/Ability." In *DisCrit: Disability Studies and Critical Race Theory in Education*. Disability, Culture, and Equity Series. New York: Teachers College Press, 2016.

Connor, Richard, Stefan Uhlenbrook, and Engin Koncagül. *Leaving No One Behind: The United Nations World Water Development Report 2019, Excutive Summary*. Perugia: UNESCO World Water Assessment Programme, 2019.

Cribb, Julian. *Food or War*. New York: Cambridge University Press, 2019.

Dave C. "KWA Pipeline- Jeff Wright." West Flint Business Association, April 23, 2015. https://web.archive.org/web/20151020061843/http://westflintbiz.com/kwa-pipeline-jeff-wright/

De La Torre, Miguel A. *Santería: The Beliefs and Rituals of a Growing Religion in America*. Grand Rapids, MI: Eerdmans, 2004.

———. *Doing Christian Ethics from the Margins*. Maryknoll, NY: Orbis Books, 2014.

———. *The Politics of Jesús: A Hispanic Political Theology*. New York: Rowman & Littlefield, 2015.

Delgado, Richard, and Stefancic, Jean. *Critical Race Theory: An Introduction*. New York: New York University Press, 2012.

Del Toral, Miguel A. *High Lead Levels in Flint, Michigan—Interim Report*. Washington, DC: United States Environmental Protection Agency—Region 5, June 24, 2015.

Detroit Program Office. *Report on Water Pollution in the Lake Huron Basin: Flint River*. Grosse Ile, MI: US Department of the Interior, Federal Water Pollution Control Administration, December 1966.

Deloria, Vine, Jr. "The Concept of History." In *Spirit and Reason: The Vine Deloria, Jr. Reader*, edited by Barbara Deloria, Kristen Foehner, and Sam Scinta. Golden: Fulcrum, 1999.

Du Bois, W. E. B. *Darkwater: Voices from Within the Veil.* New York: Harcourt, Brace and Howe, 1920.

———. *The Soul of Black Folks.* Edited by Henry Louis Gates Jr. New York: Oxford University Press, 2007.

Easterbrook, Gregg. *A Moment on the Earth: The Coming Age of Environmental Optimism.* New York: Viking, 1995.

Eck, Diana L. "GAṄGĀ: The Goddess Ganges in Hindu Sacred Geography." In *Devī: Goddesses of India,* edited by John S. Hawley and Donna M. Wulff. Delhi: Motilal Banarsidass, 1998.

Eisner, Jennifer. "MDHHS Releases October Data Outlining Blood Lead Levels in Flint." Lansing: Michigan Department of Health and Human Services, December 3, 2015.

Environment America. "Get the Lead Out." 2019. https://environmentamerica.org/feature/ame/get-lead-out.

———. "Get the Lead Out: Back-to-School Toolkit." 2019. https://environmentamerica.org/feature/ame/get-lead-out-back-school-toolkit.

Environmental Protection Bureau. "Flint River Study, August 6–7, 1974." Lansing: Michigan Department of Natural Resources, February 1977.

Estes, Nick. *Our History Is the Future: Standing Rock versus the Dakota Access Pipeline, and the Long Tradition of Indigenous Resistance.* Brooklyn: Verso, 2019.

Esteva, Gustavo, and Madhu Suri Prakash. *Grassroots Postmodernism: Remaking the Soil of Culture,* 3rd ed. London: Zed Books, 2014.

Estrine, Saul, and Adeline Pelletier. "Privatization in Developing Countries: What Are the Lessons of Recent Experience?" *The World Bank Research Observer* 33, no. 1 (February 2018): 65–102.

EWASH and Al-Haq. Joint Parallel Report to the Committee on Economic, Social and Cultural Rights on the Occasion of the Consideration of the Third Periodic Report of Israel. *Israel's Violations of the International Covenant on Economic, Social and Cultural Rights with Regards to the Human Rights to Water and Sanitation in the Occupied Palestinian Territory* (September 2011). https://www2.ohchr.org/english/bodies/cescr/docs/ngos/EWASH-Al-Haq_Israel_CESCR47.pdf.

Fedinick, Kristi, Steven Taylor, and Michele Roberts. *Watered Down Justice.* New York: National Resources Defense Council, 2019.

Flyvbjerg, Bent. *Making Social Science Matter: Why Social Inquiry Fails and How It Can Succeed Again.* Cambridge: Cambridge University Press, 2001.

Fields, Karen E., and Barbara J. Fields. *Racecraft: The Soul of Inequality in American Life.* London: Verso, 2012.

George, Susan, and Fabrizio Sabelli. *Faith and Credit: The World Bank's Secular Empire.* Boulder, CO: Westview, 1994.

Gnanadason, Aruna. "Toward a Feminist Eco-Theology for India." In *Women Healing Earth: Third World Women on Ecology, Feminism, and Religion,* edited by Rosemary Radford Ruether. Maryknoll, NY: Orbis Books, 1996.

Goodman, Nanette, Michael Morris, Kelvin Boston, and Donna Walton. *Financial Inequality: Disability, Race and Poverty in America.* Washington, DC: National Disability Institute, 2017.

Grier, Laura, Delia Mayor, and Brett Zauner. *Assessing the State of Environment Justice in Michigan*. Ann Arbor: University of Michigan School for Environmental and Sustainability in Partnership with the Michigan Environmental Justice Coalition, 2019.

Guha, Ramachandra, and Joan Martinez-Alier. *Varieties of Environmentalism: Essays North and South*. London: Earthscan, 1997.

Hameeteman, Elizabeth. *Future Water (In)Security: Facts, Figures, and Predictions*. Columbus, OH: Global Water Institute, 2013.

Hanna-Attisha, Mona, Jenny LaChance, Richard Casey Sadler, and Allison Champney Schnepp. "Elevated Blood Lead Levels in Children Associated with the Flint Drinking Water Crisis: A Spatial Analysis of Risk and Public Health Response." *American Journal of Public Health* 106, no. 2 (2016): 283–90.

Harding, Vincent. *There Is a River: The Black Struggle for Freedom in America*, reissue ed. Boston: Mariner Books, 1993.

Harris, Melanie L. *Ecowomanism: African American Women and Earth-Honoring Faiths*. Maryknoll, NY: Orbis Books, 2017.

Hartman, Saidiya. *Lose Your Mother: A Journey along the Atlantic Slave Route*. New York: Farrar, Straus and Giroux, 2008.

Hawkins, Mike. *Social Darwinism in European and American Thought, 1860–1945: Nature as Model and Nature as Threat*. New York: Cambridge University Press, 1997.

Highsmith, Andrew R. "Demolition Means Progress: Urban Renewal, Local Politics, and State-Sanctioned Ghetto Formation in Flint, Michigan." *Journal of Urban History* 35, no. 3 (March 2009): 348–68.

Hill-Landolt, Julian, Daniel Fule, and Maria Mendiluce. *2019 Outlook & Trends*. Geneva: World Business Council for Sustainable Development, 2019.

Hurley, Andrew. *Environmental Inequalities: Class, Race, and Industrial Pollution in Gary, Indiana, 1945–1980*. Chapel Hill: University of North Carolina Press, 1995.

Holmes, Barbara A. *Race and the Cosmos: An Invitation to View the World Differently*. 1st ed. Harrisburg, PA: Bloomsbury T&T Clark, 2002.

Hong, Christine J. *Identity, Youth, and Gender in the Korean American Church: Asian Christianity in the Diaspora*. New York: Palgrave Pivot, 2015.

Hutchins, Wells A. *Water Rights Law in the Nineteen Western States* 1. Clark, NJ: Lawbook Exchange, 2004.

Ira J. Taylor Library Archives. *The Iliff Family Papers Biographical Sketches*. Denver: Marshall Edison, 2002.

Johnson, Jay, Gary Pecquet, and Leon Taylor. "Potential Gains from Trade in Dirty Industries: Revisiting Lawrence Summers' Memo." *Cato Journal* 27, no. 3 (2007): 397–410.

Jo, Eung-Tae. "The 'Water Faith' of Korean New Religions." *The Korean Academy of New Religions* 22 (April 2010): 151–81.

John Paul II, Pope. *The Gospel of Life*. New York: Random House, 1995.

Joint Monitoring Programme (JMP). *Progress on Drinking Water, Sanitation and Hygiene: 2017 Update and Sustainable Development Goal Baselines*. Geneva: World Health Organization and the United Nations Children's Fund, 2017.

Kang, Nam-Soon. "Confucian Familism and Its Social/Religious Embodiment in Christianity: Reconsidering the Family Discourse from a Feminist Perspective." *Asia Journal of Theology* 18, no.1 (April 2004): 168–89.

Kang, Yong-Ki. "Environmental Ethics in Shamanism: Centering on the Korean Practice of Shamanism." *The Korean Society of East-West Comparative Literature* 20 (June 2009): 7–28.

Kelly, W. Osmund, Tom Kay, and W. Dale Bachtel. "Flint's Industrial Water Supply—Joint Discussion." *American Water Works Association* 63, no. 3 (March 1971): 148–53.

Kidwell, Clara Sue, Homer Noley, and George E. "Tink" Tinker. *A Native American Theology*. Maryknoll, NY: Orbis Books, 2001.

Kim-Cragg, HyeRan. *Interdependence: A Postcolonial Feminist Practical Theology*. Eugene, OR: Pickwick, 2018.

King, Joyce Tekahnawiiaks. "The Value of Water and the Meaning of Water Law for the Native Americans Known as the Haudenosaunee." *Cornell Journal of Law and Policy* 16, no. 3 (Summer 2007): 449–72.

King, Tiffany Lethabo. *The Black Shoals: Offshore Formations of Black and Native Studies*. Durham, NC: Duke University Press, 2019.

Knowlton, Christopher. *Cattle Kingdom: The Hidden History of the Cowboy West*. New York: Houghton Mifflin Harcourt, 2017.

Kuhn, Thomas S. *The Structure of Scientific Revolutions*. Chicago: University of Chicago Press, 2012.

Kumar, Krishan. *Prophecy and Progress: The Sociology of Industrial and Post-Industrial Society*. New York: Penguin, 1978.

Kurtz, Edward L, Emergency Manager. *Resolution to Purchase Capacity from Karegnondi Water Authority*. Resolution # 2013EM041. City of Flint, Michigan. March 29, 2013.

Lark, Theresa, and Jeff Hammer. "'The Future Is Now' Karegnondi Water Authority Breaks Ground." *MACDC Pipeline* 22, no. 3 (Fall 2013): 20–22.

Larson, Kate Clifford. *Bound for the Promised Land: Harriet Tubman, Portrait of an American Hero*. New York: Random House, 2004.

Lee, Pil-young. "Symbolism of Well in Korean Folklore." *Korean Historical Folklore Institute* 26 (March 2008): 247–88.

Long, Charles. *Significations: Signs, Symbols, and Images in the Interpretation of Religion*. Aurora, CO: Davis Group, 1999.

Lyons, Oren. "Traditional Native Perspectives." *Orion Nature Quarterly* 9, no. 3 (Summer 1990): 33–34.

Martinez-Alier, Joan. *The Environmentalism of the Poor*. United Nations Research Institute for Social Development (UNRISD) and University of Witwatersrand, August 30, 2002. http://www.unrisd.org/80256B3C005BCCF9/(httpAuxPages)/5EB03FFBDD19EA90C1257664004831BD/$file/MartinezAlier.pdf

McCauley, Erin J. "The Cumulative Probability of Arrest by Age 28 Years in the United States by Disability Status, Race/Ethnicity, and Gender." *American Journal of Public Health* 107, no. 12 (2017): 1977–81.

McKittrick, Katherine. *Demonic Grounds: Black Women and the Cartographies of Struggle*. 1st ed. Minneapolis: University of Minnesota Press, 2006.

Michigan Senate. *Flint Water Crisis: Report of the Joint Select Committee*. Chaired by Senator Jim Stamas. Lansing: Joint Select Committee on the Flint Water Public Health Emergency, October 19, 2016.

Mooney, James. "The Ghost-Dance Religion and the Sioux Outbreak of 1890." In *Fourteenth Annual Report of the Bureau of Ethnology to the Secretary of the Smithsonian Institution 1892–93*, Part 2. Washington, DC: Government Printing Office, 1896.

Morgenthau, Henry, Jr. "Address by the Honorable Henry Morgenthau, Jr. at the Inaugural Plenary Session, July 1, 1944." In *United Nations Monetary and Financial Conference, Bretton Woods, New Hampshire, July 1 to July 22, 1944, Final Act and Related Documents*. Washington, DC: US Department of State Conference Series 55.

Muhammad, Michael, E. Hill De Loney, Cassandra L. Brooks, Shervin Assari, DeWaun Robinson, and Cleopatra H. Caldwell. "'I Think That's All a Lie . . . I Think It's Genocide': Applying a Critical Race Praxis to Youth Perceptions of Flint Water Contamination." *Ethnicity & Disease* 28, no. S1 (August 8, 2018): 241.

Murdock, Esme G. "Storied with Land: 'Transitional Justice' on Indigenous Lands." *Journal of Global Ethics* 12, no. 2 (2018): 232–39.

Nabokov, Peter. *Where the Lightning Strikes: The Lives of American Indian Sacred Places*. New York: Penguin, 2006.

National Council on Disability. "Breaking the School-to-Prison Pipeline for Students with Disabilities." June 18, 2015. https://www.ncd.gov/publications/2015/06182015.

Nevarez, Leonard. "Just Wait until There's a Drought: Mediating Environmental Crises for Urban Growth." *Antipode* 28, no. 3 (1996): 246–72.

Ng, Greer Anne Wenh-In. "Asian Sociocultural Values: Oppressive and Liberating Aspects from a Woman's Perspective." In *People on the Way: Asian North Americans Discovering Christ, Culture, and Community*, edited by David Ng. Valley Forge, PA: Judson, 1996.

Noel, Thomas J., and Duane A. Smith. *Colorado: The Highest State*. 2nd ed. Boulder, CO: University Press of Colorado, 2011.

Norgaard, Richard B. *Development Betrayed: The End of Progress and a Coevolutionary Revisioning of the Future*. New York: Routledge, 1994.

Oh, Jea S. *A Postcolonial Theology of Life: Planetarity East and West*. Upland, CA: Sopher, 2011.

Oh, SaeChang. "What Is Water for Human?" *Gyeongju Society* 6 (July 2018): 66–88.

Pak, Su Yon, Unzu Lee, Jung Ha Kim, and Myung Ji Cho. *Singing the Lord's Song in a New Land: Korean American Practices of Faith*. Louisville, KY: Westminster John Knox, 2005.

Poling, James N., and HeeSun Kim. *Korean Resources for Pastoral Theology: Dance of Han, Jeong, and Salim*. Eugene, OR: Pickwick, 2012.

Preston, Christopher J. *Grounding Knowledge: Environmental Philosophy, Epistemology, and Place*. Athens: University of Georgia Press, 2003.

Pui-lan, Kwok. *Introducing Asian Feminist Theology*. Cleveland: Pilgrim, 2000.

Ray, Sarah Jaquette. *The Ecological Other: Environmental Exclusion in American Culture*. Tucson: University of Arizona Press, 2013.

Risi, Lauren. "Beyond Water Wars." *Wilson Quarterly* 43, no. 3 (Summer 2019): 1.

Rothstein, Richard. *The Color of Law: A Forgotten History of How Our Government Segregated America.* New York: Liveright, 2018.

Rumpler, John, and Laura Deehan. *Get the Lead Out: Ensuring Safe Drinking Water for Our Children at School.* Denver: Environment America Research & Policy Center, March 2019.

Ryu, Tong Shik. *The Christian Faith Encounters the Religions of Korea.* Seoul: Christian Literature Society of Korea, 2001.

Sadler, Richard Casey, and Andrew R. Highsmith. "Rethinking Tiebout: The Contribution of Political Fragmentation and Racial/Economic Segregation to the Flint Water Crisis." *Environmental Justice* 9, no. 5 (October 2016): 143–51.

Shan hai jing. *A Chinese Bestiary: Strange Creatures from the Guideways through Mountains and Seas.* Edited and translated by Richard E. Strassberg. Berkeley: University of California Press, 2002.

Shiva, Vandana. *The Violence of Green Revolution: Third World Agriculture, Ecology and Politics.* London: Zed Books, 1991.

Sicotte, Diane. "Power, Profit and Pollution: The Persistence of Environmental Injustice in a Company Town." *Human Ecology Review* 16, no. 2 (Winter 2009): 141–50.

Skeel, David. "Hauerwasian Christian Legal Theory." *Faculty Scholarship at Penn Law* 75, no. 4 (October 1, 2012): 115–33.

Stevenson, Bryan. *Just Mercy: A Story of Justice and Redemption.* New York: Spiegel & Grau, 2014.

Stockholm International Water Institute (SIWI). *World Water Week 2019: Water for Society, Including All—Overarching Conclusions.* Johanneshov: Stockholm International Water Institute, 2019.

Szulecki, Kacper, and Kirsten Westphal. "The Cardinal Sins of European Energy Policy: Nongovernance in an Uncertain Global Landscape." *Global Policy* 5 (October 2014): 38–51.

Taylor, Frederick Winslow. *The Principles of Scientific Management.* New York: Harper, 1911.

Temper, Leah, Federico Demaria, Arnim Scheidel, Daniela Del Bene, and Joan Martinez-Alier. "The Global Environmental Justice Atlas (EJAtlas): Ecological Distribution Conflicts as Forces for Sustainability." *Sustainability Science* 13, no. 3 (May 2018): 573–84.

Timoney, Kevin P. *The Peace-Athabasca Delta: Portrait of a Dynamic Ecosystem.* Edmonton: University of Alberta Press, 2013.

Tinker, George. "Spirituality, Native American Personhood, Sovereignty and Solidarity." In *Spirituality of the Third World: A Cry for Life,* edited by K. C. Abraham and B. Mbuy-Beya. Maryknoll, NY: Orbis Books, 1994.

Townes, Emilie M. *In a Blaze of Glory: Womanist Spirituality as Social Witness.* Nashville: Abingdon, 1995.

Unified Coordination Group. *Flint Rash Investigation: A Report on Findings from Case Interviews, Water Testing, and Dermatologic Screenings for Rashes That Developed or Worsened after October 16, 2015.* 2016.

United Nations World Commission on Environment and Development. *Our Common Future*. Edited by Volker Hauff. Oxford: Oxford University Press, 1987.

Urry, Lisa A., Michael L. Cain, Steven A. Wasserman, Peter V. Minorsky, Jane B. Reece, and Neil A. Campbell. *Campbell Biology*, 11th ed. New York: Pearson, 2016.

Vanderwarker, Amy. "Water and Environmental Justice." In *A Twenty-First Century US Water Policy*, edited by Juliet Christian-Smith and Peter H. Gleick. New York: Oxford University Press, 2012.

Vizenor, Gerald. *Fugitive Poses: Native American Scenes of Absence and Presence*. Lincoln: Nebraska University Press, 2000.

Walker, Alice. *Living by the Word*. San Diego: Harcourt Brace, 1981.

"Water Fights." *The Economist* 430, no. 9132 (March 2, 2019): S8.

World Economic Forum. *The Global Risks Report 2019*. 14th ed. Geneva: World Economic Forum, 2019.

Whyte, Kyle Powys. "Indigenous Food Systems, Environmental Justice, and Settler-Industrial States." In *Global Food, Global Justice: Essays on Eating under Globalization*, edited by Mary C. Rawlinson and Caleb Ward. Newcastle: Cambridge Scholars, 2015.

———. "Indigenous Experience, Environmental Justice and Settler Colonialism." In *Nature and Experience: Phenomenology and the Environment*, edited by Bryan E. Bannon. Lanham, MD: Rowman & Littlefield, 2016.

Wright, Beverly. "Race, Politics and Pollution: Environmental Justice in the Mississippi River Chemical Corridor." In *Just Sustainabilities: Development in an Unequal World*, edited by Julian Agyeman, Robert D. Bullard, and Bob Evans. New York: Earthscan, 2003.

Wright, Jeffrey. *Jeff Wright Statement on Flint City Council Approving Resolution to Join KWA*. Genesee County Drain Commissioner's Office, March 26, 2013.

Zinner, Christiana. "Water and Racialized Infrastructure in the Era of Trump." In *Faith and Resistance in the Age of Trump*, edited by Miguel A. De La Torre. Maryknoll, NY: Orbis Books, 2017.

Contributors

LORING ABEYTA is the John Wesley Iliff Senior Adjunct of Religion and Society at the Iliff School of Theology. A Templeton course award winner, she has taught graduate courses in security and environment, US foreign policy, and critical race theory. She holds degrees in anthropology, theology, international relations, and elementary education. Her current research interests examine the connections of worldview differences and sustainability.

MIGUEL A. DE LA TORRE is a scholar-activist tenured as professor of social ethics and Latinx studies at Iliff School of Theology in Denver, Colorado. Since obtaining his doctorate in 1999, he has published over thirty-five books, five of which won national awards. He served as the president of the Society of Christian Ethics in 2012, and he wrote the screenplay for the documentary *Tails of Hope and Terror* (www.trailsofhopeandterrorthemovie.com).

OSCAR DE PAZ is a doctoral candidate in the energy and environmental policy program at the Biden School of Public Policy and Administration at the University of Delaware. His research emerges at the intersection of the environment, science, technology, politics, and activism.

GARAM HAN is a doctoral student in religious education at Garrett-Evangelical Theological Seminary in Evanston, Illinois. Her research interest is Korean woman's identity formation through an ecological lens.

JENNIFER S. LEATH is the assistant professor of religion and social justice and the director of the Masters of Social Justice and Ethics at Iliff School of Theology. She is currently working on a book manuscript, *From Black to Quare and Then (to) Where: Theories of Justice and Black Sexual Ethics.* Dr. Leath is also the pastor of Campbell Chapel African Methodist Episcopal Church.

MICHELLE ANDREA MARTINEZ is a Detroit-based activist and mother of two who leads the Michigan Environmental Justice Coalition. A graduate of the University of Michigan, she has worked in environmental justice issues since 2006, struggling to advance antiracist policies through community organizing, land-based practice, and cultural arts.

VED P. NANDA is the Distinguished University Professor and Director of the Ved Nanda Center for International and Comparative Law at the University of Denver Sturm College of Law. He holds leadership positions in the global international law community, has published twenty-four books and over 225 chapters and law review articles in international and comparative law, and is a regular commentator in both the electronic and print media.

RUDOLPH P. REYES II is a PhD candidate in social ethics and Latinx studies at the University of Denver and Iliff School of Theology. He earned a BA in religion from Arizona State University and a MDiv from Harvard Divinity School. His research sits at the intersections of liberative ethics, Latinx religion, neurodiversity, and disability studies.

BILL RITTER was the 41st governor of Colorado, who previously served as the district attorney of Denver (1993–2005). During his four-year term as governor, he established Colorado as a national and international leader in clean energy by building a New Energy Economy. After leaving the Governor's Office, he founded the Center for the New Energy Economy at Colorado State University, which works with state and federal policymakers to create clean energy policy throughout the country.

ERIC C. SMITH is assistant professor of early Christianity and contemporary Christian practices at the Iliff School of Theology. He is the author of three books about the origins of Christianity and the New Testament. As an ordained minister in the Christian Church (Disciples of Christ), he has worked on issues like gun violence, gender and sexuality, environmental destruction, and racism.

Index